THE MANIPULATION
OF GENETIC SYSTEMS
IN PLANT BREEDING

THE MANIPULATION OF GENETIC SYSTEMS IN PLANT BREEDING

A ROYAL SOCIETY DISCUSSION

ORGANIZED BY
H. REES, F.R.S., R. RILEY, F.R.S.,
E. L. BREESE AND C. N. LAW

HELD ON 29 AND 30 OCTOBER 1980

LONDON
THE ROYAL SOCIETY
1981

Printed in Great Britain for the Royal Society
at the
University Press, Cambridge

ISBN 0 85403 165 0

First published in *Philosophical Transactions of the Royal Society of London*,
series B, volume 292 (no. 1062), pages 399–609

Copyright

© 1981 The Royal Society and the authors of individual papers.

It is the policy of the Royal Society not to charge any royalty for the production of a single copy of any one article made for private study or research. Requests for the copying or reprinting of any article for any other purpose should be sent to the Royal Society.

Published by the Royal Society
6 Carlton House Terrace, London SW1Y 5AG

PREFACE

During a mere few hundred years plant breeders have so directed the evolution of crop plants as to achieve dramatic – even spectacular – changes in plant form and function. The changes are the consequences of selection for higher yield and quality of food products and of medicines, for superior fibres for clothing, better timbers for fuel and building and for more attractive forms as ornament. For the most part the methods adopted by the earlier plant breeders were strictly constrained by the genetic systems of the species with which they worked. Wherever possible, however, they took advantage of the system. In sexually reproducing inbreeding species and species reproducing asexually they would initially have selected the most desirable individuals and propagated their genetically identical offspring. In outbreeding species they sought to establish the desired complex of characters in a variety by hybridization and genic recombination among selected parents. To a large degree these methods are still employed, still important. But over the years, to an increasing extent, the genetic systems of species came to be manipulated, sometimes consciously, sometimes not, to make available to the breeder a far greater range of heritable variation and to allow for a greater precision in the combining and recombining of genetic material. The changes brought about in the genetic systems as the result of plant breeding are, in many respects, comparable with the changes associated with the evolution of genetic systems in natural populations, changes whose significance to adaptation and the evolution of natural species were made abundantly clear by C. D. Darlington some 40 years ago (*The evolution of genetic systems*, Cambridge University Press, 1939). In cultivated crops, as in Nature, reproductive systems have been altered by changes in floral morphology and in gene complexes affecting incompatibility. New methods for the combining and recombining of genetic material have been exploited, methods that depend on polyploidy and on chromosome structural change. Very recently, however, possibilities for the manipulation of genetic systems have been suddenly, and to a startling degree, vastly extended. They emanate from new developments in cell biology and, particularly, in molecular genetics. This is therefore an appropriate time to review the methods and accomplishments of the plant breeder to date, and to assess the prospects of exploiting these new methods of manipulating genetic systems for the improvement of crop plants. In so doing we learn much, not only of the problems and prospects in plant breeding, but also about the many factors that affect the responses to selection upon the phenotype and about their consequences. In other words we learn a great deal not only about the application of biology but about the very fundamentals of the science.

The organizers of the Discussion Meeting, at which these papers were presented, are very grateful for the generous assistance of the staff of the Royal Society, in particular Miss C. A. Johnson and Dr M. B. Goatly. We extend to them our sincere thanks.

April 1981

H. REES
R. RILEY
E. L. BREESE
C. N. LAW

Dedication

Professor C. D. Darlington, F.R.S., who wrote the introductory paper in this book, died on 26 March 1981. It is with the greatest respect that this book is dedicated to his memory. We owe to him the very concept of genetic systems and their evolution, and, in the light of this, a clearer understanding of the consequences of their manipulation.

CONTENTS

[One plate]

	PAGE
PREFACE	[v]

C. D. DARLINGTON, F.R.S.
    Genetics and plant breeding, 1910–80   [1]

J. L. JINKS, F.R.S.
    The genetic framework of plant breeding   [7]

W. WILLIAMS
    Methods of production of new varieties   [21]
        Discussion: D. R. KNOTT   [30]

J. P. COOPER, F.R.S.
    Physiological constraints to varietal improvement   [31]

J. BINGHAM, F.R.S.
    The achievements of conventional plant breeding   [41]
        Discussion: P. S. WELLINGTON, M. S. WOLFE, D. R. JOHNSTON   [54]

R. A. NILAN
    Induced gene and chromosome mutants   [57]

A. DURRANT
    Unstable genotypes   [67]
        Discussion: H. REES, F.R.S., C. A. CULLIS   [74]

M. D. BENNETT
    Nuclear instability and its manipulation in plant breeding   [75]

E. L. BREESE, E. J. LEWIS AND G. M. EVANS
    Interspecies hybrids and polyploidy   [87]

J. G. T. HERMSEN AND M. S. RAMANNA
    Haploidy and plant breeding   [99]
        Discussion: J. HESLOP-HARRISON, F.R.S., A. P. M. DEN NIJS   [106]

C. N. LAW, J. W. SNAPE AND A. J. WORLAND
    Intraspecific chromosome manipulation   [109]

H. THOMAS
    Interspecific manipulation of chromosomes   [119]

CONTENTS

	PAGE
R. RILEY, F.R.S., C. N. LAW AND V. CHAPMAN	
The control of recombination	[129]
Discussion: R. JOHNSON	[134]
C. J. DRISCOLL	
Perspectives in chromosome manipulation	[135]
Discussion: T. E. MILLER	[145]
D. R. DAVIES	
Cell and tissue culture: potentials for plant breeding	[147]
Discussion: S. BRIGHT	[156]
E. C. COCKING	
Opportunities from the use of protoplasts	[157]
Discussion: E. THOMAS	[168]
H. REES, F.R.S., AND R. K. J. NARAYAN	
Chromosomal DNA in higher plants	[169]
Discussion: P. R. DAY	[178]
R. B. FLAVELL	
The analysis of plant genes and chromosomes by using DNA cloned in bacteria	[179]
Discussion: P. R. DAY, J. HESLOP-HARRISON, F.R.S.	[187]
J. R. POSTGATE, F.R.S., AND F. C. CANNON	
The molecular and genetic manipulation of nitrogen fixation	[189]
Discussion: G. PONTECORVO, F.R.S.	[198]
SIR KENNETH MATHER, F.R.S.	
Perspective and prospect	[201]

Genetics and plant breeding, 1910–80

By C. D. Darlington†, F.R.S.

*Botany School, South Parks Road,
Oxford OX1 3RA, U.K.*

The early successes of genetics and plant breeding and the still earlier successes of microscopy and chromosome study led to disputes, which were aggravated by lack of understanding between languages, professions and techniques. But their primary source lay in each pioneer's insistence on a uniformity of his own natural law. Bateson's exclusion of nucleus and cytoplasm was followed (in 1926) by Morgan's exclusion merely of the cytoplasm. An anti-genetic and anti-evolutionary revival was favoured by these disputes and has left its traces with us today.

The idea of a uniformity in heredity or the genetic system is once again an obstacle to understanding. For, in the practice of plant breeding, we are faced by a conflict between evidence on experimental and evolutionary time-scales. Louis de Vilmorin, Darwin and Mendel thought of this problem under the title of the 'causes of variability'. We can now recognize that the experimental or classical models of mutation and recombination of genes and chromosomes is no longer universally sufficient either for organisms or for their chromosomes. Variability in higher organisms seems to have a variety of pre-nuclear as well as nuclear foundations.

I

The great pioneers of our subject were tormented by crises of belief and uncertainty, which we need to understand in facing our own problems today. It is only today after 70 years that such understanding is coming within our reach – and may soon slip out of our reach.

Consider the crisis (on 18 July 1910) when Bateson, arriving in Merton to set up his new John Innes Horticultural Institution, received a letter from Morgan‡ in Woods Hole telling him about his 'extraordinary luck with a sex-limited case in *Drosophila*'. Morgan enclosed the corrected manuscript of an article for *Science* and offered to send Bateson stocks of his white-eyed fly. 'Drop me a line', he added, 'about what you think of the result.'

Morgan's 'result' was one that neither of the two men could at once understand. But it quickly divided them. On the one hand, Morgan learnt to see it as reconciling Mendelism with the chromosome theory. He did so with the help of Wilson, who told him of another piece of chromosome theory, known in German by the name of 'crossing over'. On the other hand, Bateson refused to learn this lesson, as he might have done from his mentors in Cambridge, Doncaster and Gregory. Instead he devoted the rest of his life to protecting the mystery of Mendelism from the rival mystery, which he privately called the 'chromosome cult'. So it was that when he returned from his last visit to Morgan in 1922 he exclaimed, in despair (again privately), that 'all my life's work has gone for nothing'.‡

It is important for us to understand the paradox of Bateson. The very fact that he had attached his idea of genetics to his own verbal framework of Mendelism (F_1, F_2, allelomorph, homozygote and heterozygote, etc.) prevented him admitting that the microscope might make this framework part of something bigger, indeed something deserving his name of genetics.

† Died 26 March 1981. ‡ See notes.

It was tragic that this happened at the moment when he was engaged in setting up his own centre of genetic research. He began by appointing W. O. Backhouse. Within 2 years this young but already experienced plant breeder had discovered the capacities and limitations of inbreeding and outbreeding in a wide range of horticultural plants. This was not Mendelism, however, and Backhouse was quickly despatched to breed wheat for the Argentine government. But from seeing the results of this work, another young man, N. I. Vavilov, who came a year later, picked up and took away his own long-range ideas about species and their systems of variation. He could distinguish polyploid wheats, $4x$ and $6x$, by their diseases before their chromosome numbers were known.

One after another the plant species and genera that Bateson was breeding revealed polyploidy, a chromosome property, that he was determined to disregard as irrelevant to heredity. Chromosomes and genes, and above all the incompatibility gene, were set on one side.

A climax came in 1926, when two opposed and partly mistaken views of heredity were reached. On the one hand, Morgan in *The theory of the gene* gave the whole of heredity to the nucleus and had nothing left for the cytoplasm. On the other hand, Bateson, because he refused to give anything to the nucleus, also could allow nothing for the cytoplasm. He found himself caught in a trap by the cross between oil and fibre flaxes, with its conflict between nucleus and cytoplasm. He extricated himself with a verbal fantasy, a piece of abstract Mendelism which he called 'anisogeny'.

Thus the role of the cytoplasm, which was evident to the German botanists, Erwin Baur, Fritz von Wettstein and Otto Renner, was hidden from our zoologists, Bateson and Morgan, and hidden also from a generation of plant breeders.

At this point, the John Innes work had comfortably split itself as though in two adjoining university departments. On the one side there was breeding of plants without looking at their chromosomes; on the other there was looking at chromosomes without breeding the plants. This could not go on for long and in 1924, innocently unaware of the rules, I began to look at the chromosomes of *Prunus*, *Pyrus* and *Rubus*, the original subjects of Backhouse's experiments. At a meeting of the Genetical Society in December 1925, with Bateson in the chair for the last time, I explained my results. Polyploidy occurred and was to be connected with the fertility and sterility found in fruit-tree breeding. I had no idea that in reaching this obvious conclusion I was exposing to the public view what had been Bateson's private nightmare.

With the death of Bateson in February 1926 the taboo in his own institution against the chromosome theory quietly crumbled. *Primula*, *Pisum* and *Campanula* as well as the fruit species soon yielded to the attack. Plant breeding could be put on diverse, verifiable and particulate foundations. Natural and artificial selection also fell into place. But, owing to Bateson's prestige, the damage that he had done to his own subject, to genetics, remained. His prejudice reinforced just those ignorant beliefs in the world outside that he himself had most strenuously condemned. Anti-evolutionism, behaviourism, Lamarckism, and later Lysenkoism and mere obscurantism, were all sustained and the ground for future conflict was prepared.

One of these conflicts concerned the time-scale or, if you like, the evolutionary horizon, proper to our kind of enquiry. In plant breeding, I would say, there is a scientific horizon some 10 000 years back and far, far, longer forward. But there is a practical horizon no longer than a man's lifetime. The difference between the two horizons was already apparent in 1923 when Bateson was invited to be Chairman of a new joint committee of the Ministry of Agriculture and the Royal Horticultural Society (a kind of hybrid or quango). Its scientific object was to 'compare' all fruits, including new products of plant breeding. Its practical object was to 'standardize' and maintain old commercial varieties of fruit.

On account of the double basis of fruit growing in this country, these two objects were inevitably, as they still are, in conflict. On one side were a few hundred commercial growers with a united economic interest in their capital investment. They sell fruit; they resist the introduction of better new varieties. On the other side were a thousand times as many private gardeners who grow fruit. They eat it themselves and they want the best. These latter have relied for 200 years on the success, a world-wide success, of enterprising nurserymen such as Veitch, Rivers and Laxton. It was they who had collaborated with Bateson and with Backhouse. They were the people who would be needed in the future as in the past for conserving the diversity of perennial fruit crops on the principles of Vavilov. Since the death of Bateson, these principles have been disregarded in favour of short-term aims. Together with the name of the Institution, the new John Innes fruit varieties have accordingly been suppressed, with economic consequences that are now painfully recognized.

II

Fifty years after Bateson and Morgan, microscopic, molecular and experimental techniques have revealed to us connections and continuities they did not and could not know. The genetic materials of plants and animals in Nature, we can see, are programmed in genetic systems that have evolved and, what is more, are evolving on an evolutionary time-scale. We are compelled to work mainly on an experimental time-scale. But the continuity of the chromosomes allows us to compare the two scales. We can therefore now ask ourselves how the chromosomes transfer from one scale to the other.

This question may be put in several ways. The information in the chromosomes represents their experience accumulated under natural selection and sometimes promoted by feedback trends over immense periods of time. Why, then, can the plant breeder, attempting to draw on this credit account, sometimes draw only a blank? Or, to put it in the forgotten terms of Louis de Vilmorin, Darwin and Mendel: what are the natural causes of differences in variability between species?

In broaching these questions, we are better equipped than our predecessors. We have our models. We have, first, a heredity, determined by the linear organization of DNA in the chromosomes of the nucleus and of the pre-nuclear plastids and mitochondria. And secondly, we have evolution, proceeding, we presume, on classical experimental assumptions, by breakage, reunion and recombination in these chromosomes, limited or promoted by natural selection.

At once, equally in the vertebrate animals and in the angiosperm plants that we know best, these models face us with four apparently non-classical and non-Darwinian questions:

(i) Why are there great evolutionary jumps in chromosome size unrelated to any apparent change in the character of the organism?

(ii) Why are there evolutionary changes in parts of chromosomes that we know to be excluded from crossing over?

(iii) Why are there sharp differences in the variability of species?

(iv) Why do such differences occur in respect of both the structure of organisms and the structure of their chromosomes, the two sometimes apparently unrelated to one another?

The last two questions are the ones of practical interest to us here. Can we now distinguish species of flowering plants of a known and comparable history that differ decisively in their variability? For comparison, they must be diploids, free from hybridization, with a common uniform origin, and with still largely uniform and interfertile chromosome complements. Among such plants we may take four examples of extremely high variability revealed and preserved by selection (table 1).

TABLE 1. FOUR HIGHLY VARIABLE CULTIVATED PLANT SPECIES COMPARED IN HISTORY, CHARACTER AND CHROMOSOMES (Darlington & Wylie 1965)

1. *Zea mays* from *Euchlaena mexicana*	$2n = 20$	America (Galinat 1971; Beadle 1975; Darlington 1973)	agricultural annuals, disruptively selected for some 5000 years: innumerable varieties
2. *Brassica oleracea*	$2n = 18$	Europe (Pease 1926; Darlington 1973)	
3. *Hyacinthus orientalis* 5000 named varieties since 1640	$2n = 16$	Syria to Europe (Darlington *et al.* 1951)	horticultural perennials consciously selected over 100–300 years
4. *Chamaecyparis lawsoniana* 200 named varieties since 1876	$2n = 22$	USA to Europe 1854 (Den Ouden *et al.* 1965)	

In all these cases there seems to have been a general preservation of interfertility among a wide range of diverse products. This is true even in *Hyacinthus*, where polyploidy has been selected but has significantly failed to affect the range of variation. Analogies in animals would be with man and perhaps his domestic animals selected for that reason.

The converse situation is that where violent chromosome changes occur without a serious alteration in the phenotype of the organisms or even their interfertility. The most obvious example is in an animal, the muntjac deer, where two species exist. They exist as closely similar kinds of organisms and in breeding they are still infertile, but one has three pairs of chromosomes (in the female) and the other has twenty-three pairs, the two having a similar total quantity of DNA (Chiarelli & Capanna 1973).

There are several directions in which we may look for explanation of these breakdowns of our classical model. One is general: perhaps the nuclear revolution failed to abolish the pre-nuclear plasmid devices of recombination whose relics therefore sporadically reappear to destabilize species. A second more specific possibility is that these relics are seen as very diverse forms in the intra-nuclear infections or transposable elements of McClintock, the genotrophic changes of Durrant, and the transformations of Pandey. Yet a third lies in the undoubted instability of heterochromatin due to the potentialities for illegitimate crossing over of highly repetitive segments (Darlington & Shaw 1959).

I suggest, however, that there are many more separate or sequential possibilities. Heredity in the higher organisms is not, as Bateson or classical genetics supposed, the expression of natural laws on inorganic principles; rather it is itself the result of evolution and is undergoing evolution. The transition from pre-nuclear to nuclear heredity is not necessarily complete or perfect or universal; genetic systems are therefore subject to many kinds of breakdown. These deserve our examination.

NOTES

(i) Morgan's letter I have placed in the John Innes Archives.

(ii) Bateson's cry of despair was uttered in the Athenaeum club to Leonard Darwin and reported at once to E. B. Ford.

(iii) The first John Innes Report seems to have been published by Bateson without permission of the John Innes Trustees. I persuaded them to reverse this rule in 1937.

BIBLIOGRAPHY (Darlington)

Backhouse, W. O. 1911 Self-sterility in plums (and other fruits). *Gdnrs' Chron.* **50**, 299.
Backhouse, W. O. 1912 The pollination of fruit trees. *Gdnrs' Chron.* **52**, 381.
Bateson, C. B. 1928 *W. B. Naturalist*. Cambridge University Press.
Bateson, W. 1911 The John Innes Horticultural Institution (Annual Report). *Gdnrs' Chron.* **49**, 179.
Bateson, W. 1913 *Problems of genetics*. Yale University Press.
Bateson, W. 1916 Experiments with flax. *J. Genet.* **5**, 199–201.
Bateson, W. 1926 Segregation (anisogeny). *J. Genet.* **16**, 201–235.
Beadle, G. W. 1975 The origin of *Zea mays*. In *The origins of agriculture* (ed. C. A. Reece), pp. 270–283. The Hague: Mouton.
Chiarelli, A. B. & Capanna, E. 1973 *Cytotaxonomy and vertebrate evolution*. London and New York: Academic Press.
Chinery, M. 1977 *The natural history of the garden* (private gardeners). London: Collins.
Chittenden, R. J. 1927 Cytoplasmic inheritance in flax. *J. Hered.* **18**, 337–343.
Chittenden, R. J. *et al.* 1927 An interpretation of anisogeny. *Nature, Lond.* **119**, 10–11.
Darlington, C. D. 1927 The behaviour of polyploids. *Nature, Lond.* **119**, 390–391.
Darlington, C. D. 1956 Natural populations and the breakdown of classical genetics. *Proc. R. Soc. Lond.* B **145**, 350–364.
Darlington, C. D. 1962 Otto Renner, 1885–1960. *Biogr. Mem. Fell. R. Soc. Lond.* **7**, 207–220.
Darlington, C. D. 1973 *Chromosome botany and the origins of cultivated plants*, 3rd edn. London: Unwin.
Darlington, C. D. 1979a Morgan's crisis. *Nature, Lond.* **278**, 786–787.
Darlington, C. D. 1979b The chromosomes as feedback systems in evolution. *Kybernetes* **8**, 275–284.
Darlington, C. D. 1980a Problems of the English apple. *Times, Lond.*, 12 August, p. 11.
Darlington, C. D. 1980b Chromosomes and organisms: the evolutionary paradoxes. In *Chromosomes today*, vol. 7, pp. 1–8. London: Unwin.
Darlington, C. D., Hair, J. B. & Hurcombe, R. 1951 The history of the garden hyacinth. *Heredity, Lond.* **5**, 233–252.
Darlington, C. D. & Shaw, G. W. 1959 Parallel polymorphism in the heterochromatin of *Trillium* species. *Heredity, Lond.* **13**, 89–121.
Darlington, C. D. & Wylie, A. P. 1965 *Chromosome atlas of flowering plants*. London: Unwin.
Den Ouden, P. *et al.* 1965 *Manual of cultivated conifers* (*Chamaecyparis*). The Hague: Nijhoff.
Durrant, A. 1962 The environmental induction of heritable changes in *Linum*. *Heredity, Lond.* **17**, 27–61.
Gairdner, A. E. 1926 *Campanula persicifolia* and its tetraploid form. *J. Genet.* **16**, 341–351.
Gairdner, A. E. 1929 Male sterility in flax: II. *J. Genet.* **21**, 117–124.
Galinat, W. C. 1971 The origin of maize. *A. Rev. Genet.* **5**, 447–478.
John Innes Horticultural Institution 1911–62 *Annual Reports*. (Publication suppressed 1912–37.)
John Innes Horticultural Institution 1929 *Conference on polyploidy*. Privately printed.
John Innes Horticultural Institution 1935 *Record of work*, 1910–35. Privately printed.
John Innes Horticultural Institution 1940 *The Fertility rules in fruit planting* (Leaflet no. 5). Privately printed.
Gregory, R. P. 1914 On the genetics of tetraploid *Primula sinensis*. *Proc. R. Soc. Lond.* B **87**, 484–492.
Gregory, R. P. *et al.* 1923 Genetics of *Primula sinensis*. *J. Genet.* **13**, 219–253.
Hay, R. 1980 New efforts to save scarce garden plants. *Times, Lond.*, 26 September, p. 20.
McClintock, B. 1951 Chromosome organization and genic expression. *Cold Spring Harb. Symp. quant. Biol.* **16**, 13–47.
Neame, B. D. 1980 An English apple a day [Letter]. *Times, Lond.*, 15 August, p. 11.
Pandey, K. K. 1980 Further evidence of egg transformation in *Nicotiana*. *Heredity, Lond.* **45**, 15–29.
Pease, M. S. 1926 Genetic studies in *Brassica oleracea*. *J. Genet.* **16**, 363–385.
Richmond, M. H. 1979 'Cells' and 'organisms' as a habitat for DNA. *Proc. R. Soc. Lond.* B **204**, 235–250.
Sansome, E. R. 1929 A chromosome ring in *Pisum*. *Nature, Lond.* **124**, 578.
Simmonds, N. W. 1962 Variability in crop plants, its use and conservation. *Biol. Rev.* **37**, 422–465.
Vavilov, N. I. 1914 Immunity to fungus diseases as a physiological test in genetics and systematics exemplified in cereals. *J. Genet.* **4**, 49–65.

The genetic framework of plant breeding

By J. L. Jinks, F.R.S.

Department of Genetics, University of Birmingham, P.O. Box 363, Birmingham B15 2TT, U.K.

The phenotypic variation that the breeder must manipulate to produce improved genotypes typically contains contributions from both heritable and non-heritable sources as well as from interactions between them. The totality of this variation can be understood only in terms of a methodology such as that of biometrical genetics – an extension of classical Mendelian genetics that retains all of its analytical, interpretative and predictive powers but only in respect of the net or summed effects of all contributing gene loci. In biometrical genetics the statistics that describe the phenotypic distributions are themselves completely described by heritable components based on the known types of gene action and interaction in combination with non-heritable components defined by the statistical properties of the experimental design.

Biometrical genetics provides a framework for investigating the genetical basis and justification for current plant breeding strategies that are typified by the production of F_1 hybrids at one extreme and recombinant inbred lines at the other. From the early generations of a cross it can extract estimates of the heritable components of the phenotypic distributions that provide all the information required to interpret the cause of F_1 heterosis and predict the properties of any generation that can subsequently be derived from the cross.

Applications to crosses in experimental and crop species show that true overdominance is not a cause of F_1 heterosis, although spurious overdominance arising from linkage disequilibria and non-allelic interactions can be. Predictions of the phenotypic distributions and ranges of recombinant inbred lines that should be extractable from these crosses are confirmed by observations made on random samples of inbred families produced from them by single seed descent. Within these samples, recombinant inbred lines superior to existing inbred lines and their F_1 hybrids are observed with the predicted frequencies.

1. Introduction

The variation that the plant breeder must manipulate to obtain superior genotypes typically contains contributions from heritable and non-heritable sources and from interactions between them. The heritable variation may exist in many forms: as differences among pure-breeding varieties or among the members of an outbreeding population, as segregants produced by inbreeding an F_1 hybrid or an outbred population and as novelties introduced by chromosome manipulation from wild ancestral material or induced by a mutagenic treatment. The non-heritable variation, on the other hand, may be generated by uncontrolled or uncontrollable heterogeneities within a glasshouse or experimental field, by differences between sites and seasons and by deliberately imposed cultural treatments.

Part of the heritable variation may be under the control of one or a few genes of sufficiently large and distinct action that their segregation and transmission from one generation to the next can be individually followed against the background of the remaining heritable and non-heritable variation by using the classical Mendelian method for discontinuous variation.

Provided that such genes retain their integrity in transmission and their independence in expression, being neither linked complexes of genes of individually smaller effect that may recombine, nor dependent in their phenotypic expression on the genetic background or the environment, they can be readily manipulated in simple and predictable ways. Their exploitation in a plant breeding programme poses neither theoretical questions for the geneticist nor practical problems for the plant breeder. They need not, therefore, concern us further here.

Typically the heritable variation is controlled by genes whose individual effects are not identifiable and together with the non-heritable variation produce a more or less continuous distribution of phenotypes. To understand fully the genetical framework of plant breeding, we therefore require a methodology such as biometrical genetics (Mather 1949) that does not depend on the identification of the phenotypic effects of individual genes but which allows us to analyse, interpret and make predictions about the totality of the variation irrespective of the number of gene loci contributing to it, the magnitudes of their individual or combined effects and the magnitude of the non-heritable and interactive sources of variation.

Biometrical genetics is an extension of classical Mendelian genetics that retains, although in a modified form, its analytical, interpretative and predictive powers, but in respect of the net or summed contributions of all gene loci. It substitutes for the characteristic segregation ratios of Mendelian genetics statistics such as mean, variance, skewness and kurtosis, which characterize the phenotypic distributions. In biometrical genetics as in Mendelian genetics we infer from observations on the phenotypes and their frequency distributions in the early generations of a breeding programme the nature of the genetic control, and from this we predict the phenotypes and their distributions in later generations. In the same way we infer complexity in this control arising from, for example, non-allelic interaction, genotype–environment interaction, maternal effects, sex linkage and linkage disequilibrium by failure of observations to agree with expectations that assume their absence and we identify the specific cause of failure by the improvement in goodness-of-fit with observations of only those expectations that allow for its presence.

The methodology of biometrical genetics (see Mather & Jinks 1971, 1977) and Jinks (1979) for comprehensive accounts) therefore provides a comprehensive framework for understanding the results of plant breeding programmes. It allows the problems to be defined, provides experimental designs and analyses for their solution and leads to interpretations in terms of the known actions and interactions of genes and hence to predictions of the outcome of future breeding and selection programmes.

2. Theory

(a) Heterosis

Whether the end product of a plant breeding programme is a recombinant inbred line or a cross between two or more such lines to produce F_1 or multiple cross hybrids, its success depends on the production of inbred lines that are superior to existing inbreds for whatever purpose they are used. Where, as is often so, these superior inbreds are sought among the segregating progenies of crosses between existing inbreds, observations made on the early generations can be used to predict the likelihood of obtaining them. The same observations can be used to analyse the genetical basis of any heterosis displayed by the F_1 hybrids and to predict the likelihood of obtaining recombinant inbred lines that will be as good as or superior to them.

The relative merits of aiming for inbreds as the better end product of an improvement programme can then be assessed, although other considerations, biological, technical and commercial, may contribute to the final choice.

The relations between the means of the phenotypic distributions of two pure-breeding lines ($\bar{P}_1$ and $\bar{P}_2$) and of their F_1 hybrid ($\bar{F}_1$) can be defined by three components, m, $[d]$ and $[h]$ as follows:

$$\bar{P}_1 = m+[d], \quad \bar{P}_2 = m-[d], \quad \bar{F}_1 = m+[h];$$

where $m = \frac{1}{2}(\bar{P}_1+\bar{P}_2)$, $[d] = \frac{1}{2}(\bar{P}_1-\bar{P}_2)$, with $\bar{P}_1 > \bar{P}_2$, and $[h] = \bar{F}_1-m$; m, defined as the mean of the phenotypic distribution of all possible pure-breeding families extractable from a cross (F_∞), is also the mid-parent value in the absence of non-allelic interactions; $[d] = r_d \sum_{i=1}^{k} d_i$ is the product of the coefficient of association–dispersion r_d, which takes values from 1 for complete association of alleles of like effect in P_1 and P_2 to zero for their complete dispersion between P_1 and P_2, and the sum of the additive genetic effects at all k loci at which P_1 and P_2 differ, $\sum_{i=1}^{k} d_i$; and $[h] = \sum_{i=1}^{k} h_i$ is the dominance deviation summed over all k loci that are heterozygous in the F_1.

Heterosis, that is F_1 superiority over its better parent (P_1 or P_2) will occur when $|[h]| > [d]$, the magnitude of the heterosis then being $[h] \mp [d]$, for positive and negative heterosis respectively. Heterosis cannot therefore occur unless there is net directional dominance, $[h] \neq 0$, i.e. unless at the majority of the k loci the dominance deviations have the same sign, and it will have its greatest value when dominance is unidirectional, the dominance deviations all having the same sign.

In the absence of non-allelic interactions, however, for heterosis to occur a further condition must be met. Either there is overdominance, $\sum_{i=1}^{k} h_i > \sum_{i=1}^{k} d_i$, or there is dispersion, $r_d < 1$, of completely or incompletely dominant alleles, $\sum_{i=1}^{k} h_i \leqslant \sum_{i=1}^{k} d_i$, in the parents. Since no pure-breeding line extractable from a cross can have a mean that falls outside the range $m \pm \sum_{i=1}^{k} d_i$ none will reach the F_1 performance of $m + \sum_{i=1}^{k} h_i$ if overdominance is the cause. If, however, dispersion is the cause, the F_1 mean will fall within the range of the pure-breeding lines and lines at one extreme of the range will equal or surpass the performance of the F_1 depending upon whether the dominance is complete ($\sum_{i=1}^{k} h_i = \sum_{i=1}^{k} d_i$) or incomplete ($\sum_{i=1}^{k} h_i < \sum_{i=1}^{k} d_i$).

Neither r_d nor $\sum_{i=1}^{k} d_i$ can be estimated from the components of the means of the phenotypic distributions. Estimates of r_d or of the relative values of $\sum_{i=1}^{k} d_i$ and $\sum_{i=1}^{k} h_i$ cannot therefore be used to distinguish between the two causes of heterosis. To do so the components of the variances of the phenotypic distributions are required.

In the absence of significant non-allelic interactions and genotype–environment interactions there are only two heritable components of variation, the additive genetic, D, and the dominance, H, and as many non-heritable components (for example E_w, E_b (Mather & Jinks 1977)) as the structure of the experimental design requires. If there is no significant linkage disequilibrium, estimates of D and H provide an estimate of the dominance ratio $\sqrt{(H/D)} = \sqrt{(\sum_{i=1}^{k} h_i^2 / \sum_{i=1}^{k} d_i^2)}$. This has the property of being zero for no dominance, between zero and unity for incomplete dominance, equal to unity for complete dominance and greater than unity for overdominance.

(b) *The range and phenotypic distributions of recombinant inbreds*

The range of the pure-breeding, recombinant lines around the mean of all such lines ($F_\infty = m$) that are extractable from a cross is $\pm \sum_{i=1}^{k} d_i$. If the dominance ratios h_i/d_i at the individual loci do not vary significantly over all loci then

$$\sum_{i=1}^{k} d_i \Big/ \sum_{i=1}^{k} h_i = \pm \sqrt{\left(\sum_{i=1}^{k} d_i^2 \Big/ \sum_{i=1}^{k} h_i^2\right)},$$

as can be shown by substituting the constant f for the ratio h_i/d_i,

$$\sum_{i=1}^{k} d_i \Big/ f \sum_{i=1}^{k} d_i = \pm \sqrt{\left(\sum_{i=1}^{k} d_i^2 \Big/ f^2 \sum_{i=1}^{k} d_i^2\right)},$$

whereupon

$$\sum_{i=1}^{k} d_i = \pm f \sum_{i=1}^{k} d_i \sqrt{\left(\sum_{i=1}^{k} d_i^2 \Big/ f^2 \sum_{i=1}^{k} d_i^2\right)},$$

which in terms of the estimates of the components of means and variances is $\pm [h]\sqrt{(D/H)}$ (Jinks & Perkins 1972).

This prediction can be more specific, and at the same time the condition that the dominance ratio should be constant over all k loci can be relaxed. The contributions of the heritable components to the statistics that describe the phenotypic distributions after n generations of inbreeding (F_n) are:

$$\text{mean of } F_n = m + (\beta_n)\,[h];$$

$$\text{variance in } F_n = (1-\beta_n)\,D + \beta_n(1-\beta_n)\,H;$$

$$\text{skewness in } F_n = -3\beta_n(1-\beta_n)\sum_{i=1}^{k} d_i^2 h_i + \beta_n(1-\beta_n)(1-2\beta_n)\sum_{i=1}^{k} h_i^3;$$

where β_n is the frequency of heterozygotes that remain after n generations of inbreeding.

The properties of the phenotypic distributions of the pure-breeding, recombinant lines extractable, for example, by single seed descent after many generations ($n = \infty$, $\beta_n = 0$) of inbreeding (or as dihaploids extractable from the F_1 of the cross) are therefore

$$\text{mean of } F_\infty = m,$$

$$\text{variance in } F_\infty = D,$$

$$\text{skewness in } F_\infty = 0.$$

The means of the pure-breeding lines are therefore normally distributed around an overall mean of m with a variance D and standard error $\sqrt{D}$. The frequency of families whose means deviate from m by any specified amount can be estimated from m and $\sqrt{D}$ by using the normal probability integral (Jinks & Pooni 1976; Pooni & Jinks 1978).

The foregoing treatment of the phenotypic distribution of a single character can be extended to the simultaneous treatment of two or more characters, irrespective of the genetic correlations between them, simply by introducing the corresponding components of covariation. For example, the joint distributions of two characters among the pure-breeding, recombinant lines derivable from a cross can be described and hence predicted in terms of bivariate normal distributions with means m_1 and m_2, variances D_1 and D_2 and covariance D_{12} (Pooni & Jinks 1978). Other extensions of the univariate and bivariate analyses can accommodate all the known effects of linkage disequilibria, non-allelic interactions, genotype–environment interactions and also, although this is rarely necessary, maternal inheritance and sex-linkage.

(c) Linkage disequilibrium

In the presence of linkage the additive genetic (D) and dominance (H) components of variation in the first segregating generation, the F_2, have the expectation (Mather & Jinks 1971)

$$D_{F_2} = \sum_{i=1}^{k} d_i^2 \overset{+C}{\underset{-R}{}} \sum^{\frac{1}{2}k(k-1)} 2(1-2p_{ij})\,d_i d_j$$

and

$$H_{F_2} = \sum_{i=1}^{k} h_i^2 + \sum^{\frac{1}{2}k(k-1)} 2(1-2p_{ij})^2 h_i h_j,$$

where p_{ij} is the recombination frequency between the ith and jth loci and the linkage term is added to $\sum_{i=1}^{k} d_i^2$ for each association–coupling (C) pair of linked loci and subtracted for each dispersion–repulsion (R) pair. With net directional dominance at the linked loci, H_{F_2} is inflated by linkage irrespective of its phase. D_{F_2}, on the other hand, is only affected if there is a linkage disequilibrium on summing over all $\frac{1}{2}k(k-1)$ pairs of loci. If there is a greater contribution from association–coupling combinations, D_{F_2} will be inflated. If, however, there is a greater contribution from dispersion–repulsion combinations D_{F_2} will be deflated. If H_{F_2} is inflated by net directional dominance at linked loci, we may infer overdominance from the dominance ratio ($\sqrt{(H_{F_2}/D_{F_2})} > 1$) where none exists. This is even more likely if D_{F_2} is simultaneously deflated by a greater contribution from repulsion combinations. And, of course, these two circumstances are most likely to occur simultaneously where there is heterosis arising from dispersion of dominant alleles that display net directional dominance (§2a). Linkage can therefore lead to heterosis being wrongly attributed to overdominance with all that this implies.

On the other hand, conclusions and predictions based upon D_{F_2} alone, e.g. prediction of the variance of F_∞ family means (§2b), are less affected by linkage because the balance of coupling to repulsion linkages rather than the total linkages determines the amount of inflation or deflation. Furthermore, with three or more linked loci, all pairs of loci can never be in repulsion – some must be in coupling – which limits the deflation from an excess of repulsion combinations. But, more importantly, the variance of the means of the pure-breeding lines (D_{F_∞}) obtained from the F_2 by single seed descent is

$$\sum_{i=1}^{k} d_i^2 \overset{+C}{\underset{-R}{}} \sum^{\frac{1}{2}k(k-1)} \frac{2(1-2p_{ij})}{1+2p_{ij}}\,d_i d_j.$$

It will therefore be inflated or deflated in the same direction although to a lesser extent than D_{F_2}.

In general, therefore, linkage will lead to conservative rather than misleading predictions of the probability of producing recombinant, pure-breeding lines that surpass the better parent or a heterotic F_1. Nevertheless, to interpret correctly the genetical basis of heterosis and to have prior knowledge that the predictions are conservative, it is important that an experimental design such as the triple test cross (Kearsey & Jinks 1968) is used, which allows linkage disequilibria to be detected and the direction and magnitude of the resulting biases estimated (Perkins & Jinks 1970).

(d) Non-allelic interactions

In the presence of non-allelic interactions the statistics describing the phenotypic distributions contain additional components defining the interactions between pairs of homozygous loci (i) pairs of heterozygous loci (l) and between a homozygous and heterozygous locus

(j) (Mather & Jinks 1971). For example, the expected magnitude of heterosis becomes $|[h]+[l]|-|[\pm d]+[i]|$, where $[l] = \sum^{\frac{1}{2}k(k-1)}l$, and $[i] = r_i \sum^{\frac{1}{2}k(k-1)}i$, with r_i ranging from $+1$ when all pairs of interacting alleles are in association in the parents (P_1 and P_2) to -1 when they are all in dispersion, being zero when the contributions of dispersed and associated combinations are equal in magnitude.

Heterosis will occur only if $[h]+[l] \neq 0$, that is, the dominance deviations and their interactions are mainly in the same direction, and either $\pm[d]+[i]$ is relatively small because of dispersion, $r_d < 1$ and $r_i < 1$, or $[h]+[l]$ is relatively large because of overdominance. Complementary gene interactions, in which $[h]$ and $[l]$ have the same signs, are more likely to lead to heterosis and to heterosis of a larger magnitude than duplicate gene interactions, in which they have opposite signs. Similarly, an excess of dispersion combinations among the interacting pair of genes, by making r_i negative, will increase the likelihood and magnitude of heterosis by making the sign of $[i]$ opposite to that of $\sum i$.

All of the non-allelic interaction components of the means of the phenotypic distributions can be readily and reliably detected and estimated in the early generations of a cross (Jinks & Jones 1958; Mather & Jinks 1971). One can also detect whether or not the interacting genes are linked and in linkage equilibrium (Jinks 1978). A linkage disequilibrium among genes displaying non-allelic interactions in the combinations that promote heterosis, i.e. an excess of the dispersion–repulsion phase and complementary interaction, will lead to an inflation of the dominance component if they go undetected (Mather & Jinks 1971).

The non-allelic interaction components of the variances can be readily detected but rarely estimated (Jinks & Perkins 1970; Perkins & Jinks 1970). In the presence of non-allelic interactions, estimates of D and H may be inflated or deflated depending on the type of interaction present, on whether the interacting alleles are mainly associated or dispersed in the parental lines and on the source of the estimates (Pooni & Jinks 1979). Where non-allelic interactions are more than a minor source of variation, the dominance ratio and predictions based upon it also become unreliable, but again the extent is dependent on the source of the estimates; those from a triple test cross are less affected than others.

The properties of the distribution of pure-breeding lines extractable from a cross can, however, still be simply defined (Jinks & Pooni 1976). The mean of F_∞ is unchanged and equals m, although the mid-parent value, $\frac{1}{2}(\bar{P}_1+\bar{P}_2)$, now equals $m+[i]$, the variance of F_∞ family means $= D+I$, where $I = \sum i^2$ and the skewness of F_∞ is no longer zero but $6\sum d_i d_j i_{ij} + 6\sum i_{ij}i_{jk}i_{ik}$ (Pooni et al. 1977).

(e) Genotype–environment interaction

Components that define the contributions of genotype–environment interactions to the statistics that describe the phenotypic distribution can be introduced into the models, analyses and predictions (Mather & Jinks 1971, 1977). The models and analyses can be considerably simplified, and their value for making simultaneous predictions over generations and environments can be correspondingly enhanced where, as is frequently found in practice, the interaction components are linear functions of the additive environmental components (Bucio Alanis et al. 1969; Jinks & Pooni 1980a). But while this allows elegant analyses and precise predictions in complex situations, the need for them can be readily circumvented; environmental sensitivity measured as some function of a genotype's phenotypic variance over environments is itself a character whose phenotypic distribution is amenable to all the analyses and predictions described so far (Jinks et al. 1977; Jinks & Pooni 1980b). The problems that genotype–environment interactions present are therefore the practical ones of defining and measuring their effects and not the theoretical ones of analysis and prediction.

3. Results

Final plant height in the cross between varieties 1 and 5 of *Nicotiana rustica* will be the primary source of illustrative examples but, to demonstrate the generality of the methodology and of some of the findings, these will be supplemented with examples involving other characters, crosses and crops. Table 1 gives the weighted least-squares estimates of the components of the means estimated from the mean final height of the parental varieties (P_1 and P_2) the

TABLE 1. COMPONENTS OF THE GENERATION MEANS FOR FINAL HEIGHT (CENTIMETRES) IN THE CROSS BETWEEN VARIETIES 1 AND 5 OF *NICOTIANA RUSTICA* ESTIMATED FROM PARENTS, F_1, F_2 AND FIRST BACKCROSSES RAISED IN 16 ENVIRONMENTS

genetic components			environment interaction components
common effects	m	112.44 ± 2.56	
additive genetic	$[d]$	7.21 ± 2.52	$b_d = 0.35 \pm 0.07$
dominance	$[h]$	14.78 ± 2.39	$b_h = 0.24 \pm 0.04$
goodness of fit	$\chi^2_{(3)}$	$0.27, p > 0.95$	

F_1, F_2 and first back-crosses (B_1 and B_2) when raised in 16 environments. The adequacy of a model containing only m, $[d]$ and $[h]$ with no non-allelic interactions is shown by the non-significant χ^2 testing its goodness-of-fit. To describe the means of each of the six generations in each of the 16 environments, the model requires three additional components: b_d, the linear environmental sensitivity of the additive gene action; b_h, the linear environmental sensitivity of the dominance (table 1); and e_j the additive environmental value (estimated from the mid-parent value) in each of the 16 environments. The mean of every generation derivable from this cross in any environment can be specified by these components, so these estimates can be used to predict the expected means. For example, in the nth generation of inbreeding in the jth environment, the expected mean

$$\bar{F}_{nj} = m + \beta_n [h] + (1 + \beta_n b_h) e_j,$$

where β_n is the frequency of heterozygotes. On substitution of the estimates (table 1),

$$\bar{F}_{nj} = 112.44 + 14.78\beta_n + (1 + 0.24\beta_n) e_j.$$

In all except the very best environments the F_1 shows heterosis, being taller than its taller parent (P_1 = variety 5). The average expected heterosis $\bar{F}_1 - \bar{P}_1 = [h] - [d] = 7.57$, that observed being 6.60, and its expected value in the jth environment is

$$\bar{F}_{1j} - \bar{P}_{1j} = [h] - [d] + (b_h - b_d) e_j,$$

which on substitution becomes

$$\bar{F}_{1j} - \bar{P}_{1j} = 7.57 + 0.11 e_j.$$

If it is known that other sources of variation are unimportant, D and H can be estimated by observing the variation (V) within the six generations, with the use of the three equations

$$E_w = V_{P_1} = V_{P_2} = V_{F_1},$$
$$D = 4V_{F_2} - 2(V_{B_1} + V_{B_2}),$$
$$H = 4(V_{B_1} + V_{B_2}) - 4V_{F_2} - 4E_w.$$

The F_2 triple test cross, in which each plant in a random sample of about 20 F_2 plants is crossed to P_1, P_2 and their F_1, provides tests of significance for additive, dominance and non-allelic interaction components of variation (Kearsey & Jinks 1968; Jinks & Perkins 1970a). Further independent tests detect genotype–environment interactions, linkage and linkage of genes that are interacting (Perkins & Jinks 1970, 1971; Jinks 1978). In most circumstances it has provided the best available estimates of D and H.

For the cross between varieties 1 and 5, the dominance ratio $\sqrt{(H/D)}$ based upon F_2 triple test cross estimates of D and H is 0.365 (Jinks & Perkins 1970). Dispersed, incompletely dominant alleles are therefore responsible for the heterosis. The triple test cross analysis showed no evidence of a non-allelic interaction component of variance, but provided clear evidence of linkage disequilibrium due to an excess of dispersion–repulsion combinations. The dominance ratio may therefore be too high (see §2c).

Pure-breeding lines taller than the heterotic F_1 should be extractable from this cross, and their expected range around the F_∞ mean of m is $\pm \sum_{i=1}^{k} d_i = \pm [h] \sqrt{(D/H)} = 14.78/0.365 = \pm 40.51$ (see §2b).

The tallest pure-breeding line should therefore have a final height of $m + \sum_{i=1}^{k} d_i = 152.95$ which, as expected, is taller than the heterotic F_1 with an expected mean of $m + [h] = 127.22$ and an observed mean of 126.19, and agrees well with the mean of the tallest F_8 family out of a sample of 82 of 156.18.

The corresponding analyses for the most heterotic cross out of 28 examined (Jinks 1954), that between varieties 2 and 12 of *Nicotiana rustica*, gave a dominance ratio of 0.89 and a prediction of 168.02 for the tallest pure-breeding line extractable from the cross. This, of course, surpasses the F_1 with a final height of 156.27 but agrees well with the tallest pure-breeding family in a sample of 60 of 171.62. There can therefore be little doubt that dispersed dominant alleles, rather than overdominance, is the cause of heterosis for final height in this species.

A typical cross section of the dominance ratios observed for contrasting characters in four widely different crops is presented in table 2. With the exception of those for maize, which are based on an NCM III analysis of F_2s (Comstock & Robinson 1952), all are based on F_2 triple test cross analyses. Among the estimates from the latter, ear emergence in the cross between barley varieties (Golden Promise and Mazurka) is alone in suggesting overdominance (W. T. B. Thomas, C. R. Tapsell & A. M. Hayter, personal communication). But in five crosses involving seven barley varieties the average dominance ratio for ear emergence is 0.81; overdominance is not, therefore, a general property of this character. Furthermore, the triple test cross analysis shows that a linkage disequilibrium arising from an excess of dispersion–repulsion combinations is present in the GP × M cross, as also is epistasis.

In maize, only yield shows apparent overdominance. This is typical. Unfortunately the NCM III design does not provide tests for the two known causes of spurious overdominance. However, re-estimates of these dominance ratios after the crosses had been subjected to many

TABLE 2. ESTIMATES OF THE DOMINANCE RATIO, $\sqrt{(H/D)}$, OBTAINED FROM INVESTIGATIONS OF F_2s BETWEEN PAIRS OF INBRED LINES FOR A RANGE OF CONTRASTING CHARACTERS AND CROPS

mating design	triple test cross					NCM III	
crop	*Nicotiana rustica*		*Nicotiana tabacum*†		barley‡	maize§	
character	cross VI × V5	V2 × V12	SCR × S3	GP × M	BH4/143/2 × AR	C121 × NC7	NC33 × K64
final height	0.36	0.89	0.68	0.86	0.76	0.72	0.95
time of flowering/ear emergence	0.53	0.70	0.39	1.31	0.17	0.79	0.70
yield	—	—	0.79	0.88	0.91	1.98	1.45

Sources of data: † Coombs (1980); ‡ Thomas, Tapsell & Hayter (personal communication); § Gardner *et al.* (1953).

TABLE 3. ESTIMATES OF THE DOMINANCE RATIO, $\sqrt{(H/D)}$, OBTAINED FROM TRIPLE TEST CROSS ANALYSES OF F_2s AND ADVANCED GENERATIONS DERIVED FROM THEM BY SINGLE-SEED DESCENT (F_7, F_{11} and F_{13})

cross	V1 × V5		V2 × V12	
character	final height	flowering time	final height	flowering time
generation				
F_2	0.36	0.53	0.89	0.70
F_7/F_{11}	0.36	0.27	0.67	0.58
F_{13}	0.36	0.31	—	—

TABLE 4. ESTIMATES FROM THREE SOURCES OF THE EXPECTED STANDARD ERROR, $\sqrt{D}$ OF THE DISTRIBUTION OF FINAL HEIGHT FOR THE MEANS OF THE PURE-BREEDING, RECOMBINANT FAMILIES (F_∞) PRODUCED FROM A CROSS OF VARIETIES 1 AND 5 OF *NICOTIANA RUSTICA* AND THE OBSERVED STANDARD ERROR FOR 82 FAMILIES OF THE F_{11}

source of estimates	estimated standard error $\sqrt{D}$
expected	
F_2, B_1 and B_2	13.86
F_2, NCM III	13.58
F_2, t.t.c.	14.07
observed	
F_{11} families	13.79

rounds of random mating to reduce any linkage disequilibrium gave values of 1.09 and 0.62, respectively (see Gardner (1963) for summary).

Equivalent comparisons have been carried out with the two *N. rustica* crosses by using triple test cross analyses of their F_2 and of advanced generations derived from them by single seed descent (Jinks & Perkins 1970; Pooni *et al.* 1978). This reduces the linkage disequilibrium bias of the dominance ratio which changes from

$$\sqrt{\left(\frac{\sum_{i=1}^{k} h_i^2 + \sum^{\frac{1}{2}k(k-1)} 2(1-2p_{ij}) h_i h_j}{\sum_{i=1}^{k} d_i^2 {}_{-R}^{+C} \sum^{\frac{1}{2}k(k-1)} 2(1-2p_{ij}) d_i d_j} \right)}$$

in the analysis of the F_2 to

$$\sqrt{\left(\frac{\sum_{i=1}^{k} h_i^2 + \sum^{\frac{1}{2}k(k-1)} \{2(1-2p_{ij})/(1+2p_{ij})\} h_i h_j}{\sum_{i=1}^{k} d_i^2 \overset{+C}{-R} \sum^{\frac{1}{2}k(k-1)} \{2(1-2p_{ij})/(1+2p_{ij})\} d_i d_j}\right)}$$

in the analysis of the advanced generations. Again, all the observed changes (table 3) are in the direction of a reduction in the dominance ratio, which is consistent with an excess of dispersion–repulsion linkages.

TABLE 5. ESTIMATES FROM DIFFERENT SOURCES OF THE EXPECTED STANDARD ERROR OF THE DISTRIBUTION OF A NUMBER OF CHARACTERS FOR THE MEANS OF THE PURE-BREEDING, RECOMBINANT FAMILIES (F_∞) PRODUCED FROM A CROSS OF VARIETIES 2 AND 12 OF *NICOTIANA RUSTICA* AND THE OBSERVED STANDARD ERRORS FOR 60 F_7/F_8 FAMILIES

character source of estimates	final height	flowering time	final height mean performance	final height environmental sensitivity	flowering time mean performance	flowering time environmental sensitivity
expected						
F_2, B_1 and B_2	17.03	0.00	—	—	—	—
F_3 families	25.59	8.18	25.26	8.00	10.59	3.55
F_2 t.t.c.	22.20	6.10	25.87	6.57	8.49	3.49
observed						
F_7/F_8 families	20.52	7.44	23.20	6.72	9.40	3.00

TABLE 6. THE NUMBER OF PURE-BREEDING FAMILIES THAT ARE PREDICTED AND OBSERVED TO FALL WITHIN THE SPECIFIED PHENOTYPIC CLASSES FOR FINAL HEIGHT IN THREE CONTRASTING CROSSES OF *NICOTIANA RUSTICA* VARIETIES

cross phenotypic class	V1 × V5 predicted	V1 × V5 observed	V2 × V12 predicted	V2 × V12 observed	B2 × B35 predicted	B2 × B35 observed
> heterotic F_1	15	22	1	3	—	—
> better parent	45	45	15	17	0	0
< heterotic F_1 or better parent	36	36	45	43	81	81
total families	81	81	60	60	81	81

This failure to find examples of 'true' overdominance is general. Extensive triple test cross analyses of *N. rustica* (Pooni et al. 1978) and *N. tabacum* (Coombs 1980) have failed to find a single example, while in barley no character out of eleven showed overdominance on average over the five crosses examined (Thomas, Tapsell & Hayter, personal communication), and only two characters in one of these crosses showed an apparent overdominance that could not be attributed to a linkage disequilibrium. Heterosis is therefore due to dispersed dominant alleles, and there is no reason why pure-breeding lines equal to or superior to the F_1 hybrids cannot be extracted from crosses, provided there are opportunities during their extraction for repulsion linkages to be broken by recombination.

The frequency with which these superior lines will appear among the inbred derivatives of a cross can be predicted from the expected phenotypic distribution in the F_∞ generation (§2b–e). In all circumstances, except where there is a linkage disequilibrium between pairs of genes displaying non-allelic interactions, the mean of the expected distribution, m, can be estimated from P_1, P_2, F_1, F_2, B_1 and B_2 families or an F_2 triple test cross set of families. Furthermore, these families provide a test for linkage disequilibrium between non-allelic interacting genes (Jinks 1978). Where most of the heritable variation is attributable to additivity, D, and dominance, H, any of the experimental design based upon the F_2 provide satisfactory estimates of $\sqrt{D}$, the expected standard error of the distribution of F_∞ families (Pooni & Jinks 1979). This is shown for final height in the cross of varieties 1 and 5 of $N.$ $rustica$ in table 4. In general, however, the estimate from the F_2 triple test cross is preferable for reasons already discussed, but in addition there is strong empirical evidence of the kind presented in table 5 that the estimate of $\sqrt{D}$ that it provides, even when the various sources of bias are making significant contributions, is the more reliable predictor of the standard error of the distribution of F_∞ family means.

TABLE 7. THE NUMBER OF PURE-BREEDING FAMILIES THAT ARE PREDICTED AND OBSERVED TO FALL WITHIN THE SPECIFIED PHENOTYPIC CLASSES IN RESPECT OF TWO CHARACTERS, MEAN PERFORMANCE AND ENVIRONMENTAL SENSITIVITY FOR FINAL HEIGHT, TAKEN SINGLY AND JOINTLY IN THE CROSS OF VARIETIES 2 AND 12 OF *NICOTIANA RUSTICA*

phenotypic classes			
mean performance	environmental sensitivity	predicted	observed F_7 families
> V12	—	20	23
< V12	—	39	36
> V2	—	51	47
< V2	—	8	7
> F_1	—	7	5
< F_1	—	52	54
—	> V12	46	45
—	< V12	13	14
—	> V2	50	47
—	< V2	9	12
—	> F_1	15	18
—	< F_1	44	41
> V12	> V12	18	20
> V12	< V2	1	2
< V2	> V12	5	8
< V2	< V2	3	3
within V2–V12 range		32	26
> F_1	> F_1	3	3
> F_1	< F_1	3	2
< F_1	> F_1	13	21
< F_1	< F_1	40	33

In tables 6 and 7, estimates of m and $\sqrt{D}$ from F_2 triple test cross analyses (Jinks & Perkins 1970a; Jinks & Pooni, unpublished; Jayasekara & Jinks 1976) have been used to predict some of the more interesting properties of the pure-breeding, recombinant families that should be extractable from $N.$ $rustica$ crosses. The corresponding observed properties of families from the F_7 or later generations of inbreeding are included for comparison. The predicted and observed frequencies (table 6) of pure-breeding families that are as tall as or taller than their parent and the F_1 if it shows heterosis are typical of the results obtained for single characters in crosses between $N.$ $rustica$ varieties. The predictions are reliable but, at the extremes of the distribution, conservative. The same reliability and conservatism are shown by the predictions, and corresponding observations for the joint properties of two aspects of final height in the V2 × V12 cross, mean performance and environmental sensitivity over two contrasting environmental treatments (Jinks & Pooni 1980b), which are summarized in table 7.

4. Conclusions

From the foregoing presentation of selected parts of the theory of biometrical genetics (§2) and some of its applications (§3), it should be clear that the generalization of Mendelian genetics that it embodies provides a framework within which the strategies and the results of plant breeding can be defined, analysed and understood. The theory can accommodate any level of complexity and the latest experimental designs allow the most complex situations to be resolved. As a result, true overdominance has been shown to have no demonstrable role in hybrid vigour, although spurious overdominance arising from a linkage disequilibrium with or without superimposed non-allelic interactions does. The choice between hybrids and recombinant inbred lines as the end-product of a breeding programme must therefore rest on other non-genetic considerations.

References (Jinks)

Bucio Alanis, L., Perkins, J. M. & Jinks, J. L. 1969 Environmental and genotype-environmental components of variability. V. Segregating generations. *Heredity, Lond.* **24**, 115–127.

Comstock, R. E. & Robinson, H. F. 1952 Estimation of average dominance of genes. In *Heterosis* (ed. J. W. Gowen), pp. 494–516. Ames: Iowa State College Press.

Coombs, D. 1980 Biometrical genetics of tobacco. Ph.D. thesis, Birmingham University.

Gardner, C. O. 1963 Estimates of genetic parameters in cross-fertilizing plants and their implications in plant breeding. In *Statistical genetics and plant breeding* (ed. W. D. Hanson & H. F. Robinson), pp. 225–252. Washington: National Academy of Sciences.

Gardner, C. O., Harvey, P. H., Comstock, R. E. & Robinson, H. F. 1953 Dominance of genes controlling quantitative characters in maize. *Agron. J.* **45**, 186–191.

Jayasekara, N. E. M. & Jinks, J. L. 1976 Effect of gene dispersion on estimates of components of generation means and variances. *Heredity, Lond.* **36**, 31–40.

Jinks, J. L. 1954 The analysis of continuous variation in a diallel cross of *Nicotiana rustica* varieties. *Genetics, Princeton* **39**, 767–788.

Jinks, J. L. 1978 Unambiguous test for linkage of genes displaying non-allelic interactions for a metrical trait. *Heredity, Lond.* **40**, 171–173.

Jinks, J. L. 1979 The biometrical approach to quantitative variation. In *Quantitative genetic variation* (ed. J. N. Thompson, Jr & J. M. Thoday), pp. 81–109. New York: Academic Press.

Jinks J. L., Jayasekara, N. E. M. & Boughey, H. 1977 Joint selection for both extremes of mean performance and environmental sensitivity to a macroenvironmental variable. II. Single seed descent. *Heredity, Lond.* **39**, 345–355.

Jinks, J. L. & Jones, R. M. 1958 Estimation of the components of heterosis. *Genetics, Princeton* **43**, 223–234.

Jinks, J. L. & Perkins, J. M. 1970 A general method for the detection of additive, dominance and epistatic components of variation. III. F_2 and backcross populations. *Heredity, Lond.* **25**, 419–429.

Jinks, J. L. & Perkins, J. M. 1972 Predicting the range of inbred lines. *Heredity, Lond.* **28**, 399–403.

Jinks, J. L. & Pooni, H. S. 1976 Predicting the properties of recombinant inbred lines derived by single seed descent. *Heredity, Lond.* **36**, 253–266.

Jinks, J. L. & Pooni, H. S. 1980a Non-linear genotype × environment interactions arising from response thresholds. I. Parents, F_1s and selections. *Heredity, Lond.* **43**, 57–70.

Jinks, J. L. & Pooni, H. S. 1980b Comparing predictions of mean performance and environmental sensitivity of recombinant inbred lines based upon F_3 and triple test cross families. *Heredity, Lond.* **45**. (In the press.)

Kearsey, M. J. & Jinks, J. L. 1968 A general method of detecting additive, dominance and epistatic variation for metrical traits. I. Theory. *Heredity, Lond.* **23**, 403–409.

Mather, K. 1949 *Biometrical genetics*, 1st edn. London: Methuen.

Mather, K. & Jinks, J. L. 1971 *Biometrical genetics*, 2nd edn. London: Chapman & Hall.
Mather, K. & Jinks, J. L. 1977 *Introduction to biometrical genetics*. London: Chapman & Hall.
Perkins, J. M. & Jinks, J. L. 1970 Detection and estimation of genotype–environmental, linkage and epistatic components of variation for a metrical trait. *Heredity, Lond.* **25**, 157–177.
Perkins, J. M. & Jinks, J. L. 1971 Analysis of genotype × environment interaction in triple test cross data. *Heredity, Lond.* **26**, 203–209.
Pooni, H. S. & Jinks, J. L. 1978 Predicting the properties of recombinant inbred lines derived by single seed descent for two or more characters simultaneously. *Heredity, Lond.* **40**, 349–361.
Pooni, H. S. & Jinks, J. L. 1979 Sources and biases of the predictors of the properties of recombinant inbreds produced by single seed descent. *Heredity, Lond.* **42**, 41–48.
Pooni, H. S., Jinks, J. L. & Cornish, M. A. 1977 The causes and consequences of non-normality in predicting the properties of recombinant inbred lines. *Heredity, Lond.* **38**, 329–338.

Methods of production of new varieties

By W. Williams

*Department of Agricultural Botany, University of Reading,
Whiteknights, Reading RG6 2AS, U.K.*

A review is given of the methods in general use in the breeding of crops belonging to the major groups of crop plants and an assessment is given of the important factors that determine their efficiency, especially gene–environment interaction in the early segregating generations. Variations of the standard techniques including generating homozygotes through single seed descent, cytological production of homozygous diploids, multiplication of advanced segregating bulk populations without selection, and the production of F_1 hybrid varieties in inbreeding crops, are outlined and assessed. The importance of production-scale testing to supplement data from plot trials, and its relevance to the organization of plant breeding and to the successful identification of superior selections, are considered.

It is now well understood that the prime determinant of the methods practised in plant breeding, as distinct from techniques for the induction of agronomically useful variations, is the crop's breeding system. Consideration of breeding systems focuses on the lineal relationship between the gametes at reproduction and their control of the genetic structure of populations. Of the major world crops, three of the five major cereals and most of the important oil and protein crops, including cotton, soybean and the pulses, accounting for two-thirds of the world's food, are inbreeders. Among naturally outbreeding crops, maize is pre-eminent, contributing 20% of annual grain supplies and a substantial amount of forage production; others include sunflower and many forage grasses and forage legumes. Clonal crops such as the orchard and soft fruits, sugar cane and the potato are also outbreeders in which sexual reproduction is bypassed by natural or managed methods of vegetative reproduction.

Pure-line varieties

The genetic improvement of most inbreeding crop plants is commonly operated through the development of homozygous lines automatically generated by automatic self-pollination according to routine methods of pedigree selection (Bingham 1975). This method owes its worldwide success to a few very simple, decisive factors: it is simple to understand and operate and has been widely applied even where a minimal level of support technology is available; selection of one parent is invariably predetermined by the genotypic uniqueness of the best modern varieties while the other is selected on the basis of complementing characters or on the results of diallel test crosses; the system is self-propelling, generating improved parents for subsequent hybridizations and therefore guaranteeing stepwise improvement of performance without requiring astronomically large segregating populations and, given reasonable control of mutational changes and of outcrossing during multiplication, a stable agronomic performance of selections is guaranteed.

In common with several other routine methods of plant breeding, pedigree selection is a long process from when parents are hybridized to the release of a new variety; it is therefore less responsive to constantly changing objectives than rates of change in agricultural technology frequently demand. To overcome this limitation, winter nurseries in the Southern Hemisphere, enabling two generations of selection each year up to F_5 or F_6, are now in common use, thus reducing the timescale of selection by about 25%.

The most recalcitrant problem of pedigree selection is the failure to identify genotypic values of individuals in early generations when gene–environment interactions cannot be identified, and selection decisions on economic characters like grain yield are largely an act of faith. Although significant correlations exist between performance in the early and later generations (Lupton & Whitehouse 1957; De Pauw & Shebeski 1973), the likelihood of identifying extreme genotypes in the early generations equates with random selection (Knott & Kumar 1975), which explains the irreducible level of uncertainty that still surrounds plant breeding programmes.

Variations of the pedigree selection method

Variations of the pedigree selection method have been proposed to reduce environmental effects during early selection. One of the best known involves the manipulation of the generations as bulks until F_6–F_8, subject only to natural selection and the elimination of the obviously unacceptable segregates, thus ensuring maximal retention of recombinants before selection begins. The advantage of the method as compared with low selection intensity of single plants in F_2 or F_3 is marginal, since the inevitable ear/row selections from bulks are subject to the same cryptic environmental interactions as are selections during early generations. There may be some improvement in the reliability of selecting in later generations from bulks, since homozygosity will ensure greater correspondence between genotypic and phenotypic values, but the increase in precision of selection cannot be expected to be of a very high order.

Bulk multiplication of segregating generations has also been advocated for subjecting poorly adapted genotypes to natural selection before screening for agronomic characters. Except that high density plantings allow inter-plant competition and favour aggressive genotypes, it is not clear why similar pressures do not operate in standard pedigree selection. Furthermore, natural selection and breeding objectives are seldom co-directional for economic characters, and genotypes with unacceptable expressions for characters such as tiller and seed number are liable to differential multiplication under natural selection, while others specified as important breeding objectives would be reduced or even eliminated.

It is not possible to report precisely on the use of bulk methods. It appears that the systems commonly applied are a modification of the original scheme whereby selection is postponed only during F_2 and F_3 when genotypic values are most uncertain.

Another modification of pedigree selection that is receiving attention is that known as single seed descent (see also Jinks, this symposium). In this, individual lines up to F_6 or F_8 are successively derived through a single, unselected seed saved in each generation. Large populations are grown under controlled conditions to promote reduced generation time, and homozygosis is achieved quickly, before selection commences, giving the advantage of improved correspondence of genotype and phenotype expected from delayed selection on bulk populations.

The probability of retaining the best genotypes through single seed descent equals that in pedigree selection (Knott & Kumar 1975), while theoretically it appears to allow prediction in

F_2 of the potential range of variation, thus enabling reliable culling at an early stage (Jinks & Pooni 1976). In common with the bulking of the early generations, single seed descent only partly eliminates uncertainties due to gene–environment interactions, and the method neither reduces the timescale nor the extent of field selection. Its use will necessarily be confined strictly to annual crops and its general value will depend entirely on possible advantages in retaining superior recombinants until environmental effects can be better assessed.

Another possible refinement of pedigree selection is the isolation of haploids from F_1 hybrids for the single-step creation of homozygous diploids. Two techniques are being investigated: the culture of haploid microspores (Sunderland 1978) and the formation of haploids after differential chromosome loss in progeny of F_1 hybrids. The culture of microspores has been claimed to have been successful in *Nicotiana tabaccum*, while only the cross between hybrids of *Hordeum vulgare* and *H. bulbosum* are being exploited to give haploids through chromosome loss in the F_1. Although programmes using these techniques are now established, they are still in the experimental phase, and their adaptation for routine breeding, while appearing not particularly promising, cannot yet be confidently assessed.

The importance of gene-environment interactions

Gene–environment interactions are, of course, not confined to the early segregating generations: they intrude universally throughout all programmes. Analysis of interactions indicate that although site effects are important, the larger and most refractory interactions are invariably associated with seasons. Site interactions can be minimized by careful selection of sites and by good trials management, whereas seasonal effects are unpredictable and irreducible. Thus plant breeders have no option but to select for stable performances over several seasons and therefore the long operational timescales associated with standard breeding methods are not without advantages in eliminating types that are excessively sensitive to seasonal effects.

Site interactions raise other issues, and while they are consistently smaller than seasonal effects, it is doubtful whether they are given adequate consideration in planning breeding programmes. In the famous experiments of Finlay & Wilkinson (1963), mean yields showed that, while the range in environmental sensitivity was wide, varieties with greatest stability also had highest mean yields. This pattern, however, does not exclude specific interactions leading to superior local adaptations. Extensive data from national trials by the National Institute of Agricultural Botany, Cambridge, and from the worldwide wheat variety trials conducted by the University of Nebraska indicate that site interactions are not common except across relatively very wide geographical areas. Some narrower regional adaptations are, however, indisputable. In the U.K., for example, different varieties have to be recommended for the west of England, especially to provide different patterns of disease resistance, and commonly for Scotland, where temperature differences are important. The relatively low frequency of site interactions may, however, be partly a reflection of their elimination during selection rather than a true indicator of the absence of regional adaptation: genotypes with significant interactions are likely to be discarded early during a breeding programme, whereas those exhibiting homoeostasis over sites are retained. Such influences can be minimized only by early testing at many sites and ensuring the selection of stable, contrasting patterns of adaptation, before entry into national trials.

F_1 HYBRID VARIETIES

Improvement in outbreeding species has been dominated by the highly successful programme on hybrid maize in the United States (Sprague & Eberhart 1977). The value of F_1 hybrid varieties and suitable methods for hybrid seed production had become clear in maize by 1920, and currently grain production based on hybrids predominates in every major maize-growing area of the world. Furthermore, the methodology for producing hybrid maize varieties has been applied with only minor modifications to most major outbreeding crops which allow pollination control. The crop species in which F_1 hybrids are in general use include several ornamentals, many vegetables, particularly the *Brassica* crops, onions and to a lesser extent carrots, sugar beet, and two major cereals, sorghum and maize. It is now accepted that the economic advantages of hybrids are not confined to yield alone but are also due to the impressive level of crop and product uniformity, which are invaluable features of F_1 varieties.

Outbreeding crops

The hybrid varieties initially developed in maize were genetically heterogeneous populations derived from four inbred lines, the so-called double-cross hybrids. These were a compromise between maximizing heterozygosity and the constraints on cost of hybrid seed. The greater yield and uniformity of homogeneously heterozygous hybrids based on two inbreds have only been exploited recently after the development of inbred lines capable of satisfactory seed production in single-cross hybrids. Currently, most F_1 hybrid varieties in use in the U.S.A. are single-crosses.

Following the discovery of the nucleo-cytoplasmic system of male sterility, male-sterile inbred lines whose fertility could be restored by complementary alleles completely replaced monoecious inbreds in the production of commercial F_1 seed. The superiority of the T (Texas) cytoplasm in maximizing male-sterility and in promoting high fertility restoration with appropriate restorer alleles, and its minimal pleiotropic effects on production characters, resulted in its almost universal use in female inbreds until 1970, when lines carrying this cytoplasm proved to be highly susceptible to southern corn blight and had to be discontinued. Since then a satisfactory replacement has not been developed and hybrid maize is once more being produced from mechanically detasselled, monoecious lines. This, however, must be considered to be only a temporary reversal in the genetical control of hybrid seed production in the crop.

Inbreds possessing superior combining ability are selected according to one of several well established methods. Genotypes that possess above-average combining ability can be identified during early generations of inbreeding, and since specific and general combining ability are highly correlated, initial screening is for high levels of general combining ability by using testers with a broad genetic base such as varieties ('top' crosses) or double-cross hybrids. Promising selections based on the initial test crosses are further mated to inbred testers for detecting specific combining ability in pairs of lines for commercial hybrid seed production.

GENETIC DIVERSITY AND THE ISOLATION OF COMPLEMENTARY INBREDS

Undoubtedly one of the most consistent conclusions that has emerged from the well documented hybrid maize programmes is the importance of maximizing genetic diversity between the source populations from which inbred lines are derived. The superior hybrids of the early period in the development of hybrid maize owed their success to the diversity inherent in the historical introgression of Southern Dent and Northern Eastern Flint genotypes. More recently, the same principle has emerged from hybrid maize programmes in the tropics, where production has been dramatically advanced after identification of superior combining abilities in inbreds derived from one dent and three unrelated flint populations. Inbred lines derived from these diverse stocks have generated hybrids with outstanding agronomic performances, which have quickly replaced the existing non-hybrid varieties adapted to tropical regions (Wellhausen 1978).

It has also become clear that superior inbreds can only be established after intense selection for agronomic characters, as well as rigorous screening for combining ability for grain yield. It is not surprising, given the decisive role of the source of origin of the base populations and the unpredictability of complementarity for yield between genotypes, that the frequency of isolation of superior inbred lines has been universally of the low order of less than 0.1 % of the total number studied. Clearly, the derivation of inbred lines for the production of F_1 hybrids cannot be regarded as a short-cut to success in plant breeding.

Continuing improvement of inbred lines

An obvious method to advance hybrid performance is through the recovery of improved inbreds from among selfed progeny of the best hybrids. The method is essentially pedigree selection and is applicable to outbreeding species only if self-incompatibility mechanisms do not prevent or seriously restrict inbreeding. Recombination among inbreds derived from superior F_1 hybrids allows further accumulation of alleles favourable for heterosis and second-cycle inbreds derived in this way are now in general use in commercial hybrids used in the United States. It is significant that, over the past 50 years, the ratio of the yield of inbreds to hybrids has remained stable at slightly below 0.5 (D. N. Duvick, personal communication). Thus, cycles of reselection for favourable alleles for yield improvement have failed to raise the yield of inbreds relative to hybrids, in spite of evidence for the overwhelming importance of dominance and of additive gene action in controlling yield expression.

Population improvement

The genetic constitution of the base material determines success in selection for combining ability, and its importance is reflected in extensive searches for the most efficient methods for upgrading the genetic status of populations. The objective of population improvement is simply to increase the frequency of favourable alleles before the isolation of inbred lines, and thus increase the probability of detecting superior genotypes. The methods used, which comprise both mass selection and various forms of progeny testing, are routines of varying levels of precision for accumulating alleles through recurrent selection over generations. While reflecting flexibility in respect of detail, all are designed for stepwise advances from base populations and, concurrently, to allow the production of interim hybrids, thereby reducing the time required for initial commercial exploitation. The average gain per cycle achieved by the different methods on a great range of material indicates that they are all almost equally effective, and the case for elaborate progeny tests has not been established.

Among several systems of recurrent selection, of special interest is selection within families replicated at several locations to minimize gene–environment interaction and the reconstitution of advanced generations from parents with high scores over several sites (Longquist 1964).

Reciprocal recurrent selection designed to improve the success of interpopulation crosses operates on two base populations. Reciprocal crosses, which form the basis of each selection cycle, involve tests for combining ability by using heterozygous or inbred testers from each population in combinations with selected genotypes from the other. Phenotypic selection within the two populations increases the frequency of favourable alleles for agronomic performance, while the use of progeny tests increases the probability of selecting inbreds with improved general and specific, interpopulation combining ability. A full summary of the impressive gains due to reciprocal recurrent selection for improvement in the performance of interpopulation crosses is given by Sprague & Eberhart (1977).

Incompatibility mechanisms and hybrid varieties

The development of hybrid varieties is feasible only where simple procedures for mechanical pollination control is possible as in monoecious maize, or where nucleo-cytoplasmic systems of control of male sterility allow large-scale production from male steriles as in maize, sorghum, sugar beet, onions and carrot. Exceptionally, as in *Brassica oleracea* (Thompson 1964), specialized control by the sporophytically controlled incompatibility mechanism has also been elegantly exploited.

Hybrid varieties in inbreeding crops

Of the crops that are predominantly inbreeding, the tomato is unique in that costs of hand emasculation and pollination for the production of F_1 varieties can be borne by charges on hybrid seed: in northern Europe, glasshouse tomato production is now based almost exclusively on hybrid varieties. Their success was not to be expected, since the level of heterosis recorded experimentally in the tomato has been minimal (Williams & Gilbert 1958), and pure-line varieties seemed to offer several advantages. It now seems probable that the greater opportunity for assessing gene–environment interactions in F_1 hybrids, which allow adequate replication in the first (and only) generation, may account for the undoubted success of F_1 hybrid tomatoes. Failure to identify reliable selections in F_2 and F_3 due to exceptionally large environmental interactions is probably even more important in fresh fruits and vegetables than in most other crops.

Although worldwide interest in F_1 hybrid cereals emerged after the discovery of nucleo-cytoplasmic control of sterility in the Triticineae, the exploitation of hybrid cereals has so far not proceeded generally beyond the research phase (Sage 1976; Hughes & Bodden 1978). A few hybrid varieties of barley, in which elimination of gametes containing the male fertile alleles is controlled by tertiary trisomy have, however, been developed to a commercial stage (Ramage 1965). Mutations (*ge*) closely linked to the male-fertile, *Ms*, allele, which cause the elimination of the male gametes (*Msge*), formed by male-fertile genotypes ($\frac{Msge}{msGe}$) are also being studied for the development of hybrid barley (Foster *et al.* 1979). These genotypes are non-restorers of male fertility and are essential for large-scale multiplication of ($\frac{msGe}{msGe}$) male-sterile lines for commercial seed production. The use of nucleo-cytoplasmic male sterility has recently also become possible for the development of hybrid barley following the discovery of sterile and restorer genotypes in *Hordeum spontaneum* (C. A. Foster, personal communication).

A continuing advance and the imminent release of hybrid varieties of wheat by using nucleo-cytoplasmic sterility is being claimed by seed companies in the U.S.A., but a similar programme at the Plant Breeding Institute, Cambridge, has been discontinued.

Apart from evidence that heterosis in inbreeding cereals is not large and that it can be fixed in pure lines, the restructuring of near autogamous floral mechanisms for open pollination is a major problem. Furthermore, most nucleo-cytoplasmic systems, even those extensively studied in maize, give only patchy fertility restoration in hybrids, which reinforces doubts on the potential

METHODS OF PRODUCTION OF NEW VARIETIES 427

of hybrid cereals for the near future. It may be noted, however, that F_1 hybrid seed produced by hand emasculation is reported to be developed in cotton in India where labour costs are presently not limiting, but the scale of the operation is unknown (Davies 1979).

SYNTHETIC VARIETIES

Outbreeding species that do not offer satisfactory mechanical or biological methods for developing F_1 hybrids are improved by selection of foundation genotypes for the establishment of genetically closed (isolated), open pollinating populations. Crops improved by these methods include self-incompatible forage grasses and legumes and many outbreeding vegetables. The methods do not differ significantly from population improvement and the derivation of inbreds already described, and comprise phenotypic recurrent selection as in forage grasses (Breese & Hayward 1972), or recurrent selection for combining ability based on progeny tests.

Tests for combining ability, where applied to the selection of foundation genotypes, concentrate on general combining ability by using heterogeneous testers, or, when the number of lines have been reduced to manageable numbers, polycross tests; specific combining ability between pairs of genotypes is of less importance for these varieties. The choice of whether inbred lines, heterozygous family lines or heterozygous clones are used as foundation genotypes rests on the ease of imposing inbreeding and on the vigour of inbreds in a given species. Since many outbreeding species are relatively self-incompatible, restricting the use of inbreds, heterozygous genotypes increased clonally for initial seed multiplication frequently replace sexual lines as foundation stocks.

Since synthetic varieties are maintained as closed populations during several generations of seed multiplication before commercial use, a major problem is the balance between maximizing response to selection and the prevention of loss of vigour through inbreeding, which will negate the responses. Estimates based on experimental data in maize (Kinman & Sprague 1954) indicate that six is the optimum number of unrelated foundation genotypes for maximizing advance under selection without incurring substantial reduction in vigour from inbreeding in synthetic varieties. One suspects, however, that in practice the number used is considerably greater.

CLONALLY PROPAGATED CROPS

Many crops, including citrus, most temperate orchard fruits (apples, pears, peaches and cherries), the potato, sugar cane and even some tropical forage grasses, are propagated vegetatively. In these, commercial varieties are heterozygous genotypes in which all the units of production form a genetically uniform clone except for variation due to accumulated mutations in somatic cells protected from selection. Indeed, the commercial exploitation of such spontaneous, somatic mutations has played an important role in the improvement of crops in this group.

For obvious reasons, vegetatively propagated crops are cultivated for their fruits, roots, tubers, stems or leaves, and seedlessness is frequently an additional virtue. With a few exceptions, clonal crops are outbreeders in which, *ipso facto*, production units are single heterozygotes selected from among heterogeneous family progenies. Furthermore, since many are fresh food crops, varietal specification must conform very precisely over a range of characters, while methods of breeding reflect the fundamentally different biological and commercial specifications of the various species within the group.

Selection routines in the potato

The general principles applicable to crops, which are naturally propagated vegetatively as distinct from those in which vegetative systems are imposed, e.g. apples or cherries, can be illustrated by reference to an example taken from potato breeding (Howard *et al.* 1977).

Choice of parents for hybridization was by reference to the best currently established varieties possessing complementary characters. The range of variation expressed in F_1 progeny, among which selection is practised, is particularly difficult to predict in this group of crops, and test crosses involving modest numbers of segregates are commonly studied initially to provide a guide to the total numbers required to allow a reasonable chance of recovering suitable segregates. Unpromising crosses are eliminated after preliminary test crosses, while others are repeated on a larger scale. Because of the number of character specifications that must be met through selection and of the heterozygous nature of the parents, very large progeny numbers are essential (Williams 1959).

The total number of progeny seedlings studied in the potato family cited here totalled 17582, which is average for current potato breeding programmes: examples of progeny numbers exceeding this by a wide margin are, however, common. Following elimination first of seedlings in the F_1, and at later stages of clones, on the basis of minimal criteria for seven economic characters, on general agronomic performance and on the basis of preliminary yield trials, six clones (*ca.* 0.035 % of the initial segregating population) survived for inclusion in national trials. This very modest recovery rate will almost certainly have been reduced further, probably to no more than 1 in 17000, when the results of national trials have been completed.

Intensity of selection in this programme was uniform and relatively weak over the ten clonal and one seedling generations: only segregates with obviously unacceptable environmental interactions were eliminated in any one season. As with most potato breeding programmes, selection was practised on a single site that could favour specific site interactions and lead to unsatisfactory performance in national trials where high average performance over many sites is sought. Studies on variety–site interactions in the potato have shown that major varieties such as King Edward, Maris Piper, Désirée and Pentland Crown, all show significant adaptation to certain production regions in England, indicating that site interactions should be considered in the organization of potato breeding programmes in this country.

Selection in tree crops

A number of clonally propagated crops, especially tree species, present unique problems in breeding methodology. After germination, seedlings enter a juvenile phase which may persist for several years, during which plants lack physiological competence to flower, and selection for fruit characters cannot commence. Even selection for vegetative characters such as tree form, leaf production, bud dormancy and disease resistance cannot be practised with confidence because of the general morphogenetic differences that are associated with juvenility. A reduction in juvenility is possible in some species by propagating seedlings on special root stocks, but the additional routine of handling large numbers of unselected seedlings involves substantial extra time and expense.

Since natural systems of vegetative reproduction are absent in most tree crops, selected genotypes for replicated orchard trials have to be propagated on suitable stocks. The number of years required for several cycles of vegetative reproduction places further serious limitations on the operation of plant breeding in this group, which is reflected in the overall modest success recorded in the production of much-needed new varieties in several of the crops.

The organization of plant breeding

Plant breeding in many countries is shared between private companies and government institutions. In the U.K., government institutions, notably institutes of the Agricultural Research Council and of the Departments of Agriculture, have historically played a predominant role both in variety production and in research support. Where cost/benefit ratios have been estimated (Simmonds 1974), investment in plant breeding has shown favourable returns and the involvement of private companies has increased in the U.K. during recent years since the provision of legal protection of new varieties in 1964. Currently, the size of breeding programmes in the public and private sector is judged to be approximately equal. In other countries in the European Economic Community, private companies have always played major roles in variety production, with government institutions supplying research support. This, too, is now the position in the U.S.A., where large, internationally based seed companies have become dominant in practical plant breeding.

Worldwide experience has shown that, while operational success in variety production is not a monopoly of either the public or the private sector and that adequate research support is essential for the success of both sectors, the range and complexity of problems currently associated with intensive production systems are on a scale that is new to agricultural research. Of these, the vulnerability of the major crops after erosion of nuclear and cytoplasmic variability, and the ecological imbalance created between crops and their diseases, are only the most obvious areas of concern requiring medium-term, if not long-term, perspectives in research.

Whereas the private sector will continue to depend for research support in plant breeding on the resources of the Government research services, it is clear that private breeders are frequently well placed to assess the economic worth of new selections. Many private organizations are engaged in commercial farming or horticulture on a large scale and are able to assess the behaviour of new varieties at production levels. Many crucial factors, which are difficult to quantify experimentally and which often become apparent only on a 'field' scale, can then be assessed. This is particularly relevant to crops for which adequate trial systems have not been available, and in this country it is the glasshouse, fruit and vegetable crops that have been least adequately served in this respect. It is vital, for success from investment in plant breeding, to appreciate the importance of limiting factors imposed during production and marketing, which intervene after the packets of breeder's seed have been sealed. When these practical requirements are ignored, even the most infallible genetical methodology at nursery levels will be largely wasted.

Special thanks are due to Dr D. N. Duvick, Pioneer Hibred International, Iowa, for details of current developments in maize breeding, Dr C. A. Foster, Welsh Plant Breeding Station, Aberystwyth, for a summary of advances in the production of hybrid cereals, and Mr B. D. Dowker, National Vegetable Research Station, Wellesbourne, for access to publications with restricted circulation on vegetable breeding.

References (Williams)

Bingham, J. 1975 Winter wheat breeding methods and prospects. *Jl R. agric. Soc. Engl.* **136**, 65–67.

Breese, E. L. & Hayward, M. D. 1972 The genetic basis of present breeding methods in forage crops. *Euphytica* **21**, 324–336.

Davies, D. D. 1979 Hybrid cotton: specific problems and potential. *Adv. Agron.* **30**, 129–157.

De Pauw, R. M. & Shebeski, L. H. 1973 An evaluation of early generation yield testing procedure in *Triticum aestivum*. *Can. J. Pl. Sci.* **53**, 465–470.

Finlay, K. W. & Wilkinson, G. N. 1963 The analysis of adaptation in a plant breeding programme. *J. agric. Res.* **14**, 742–754.

Foster, C. A., Fothergill, M., Hale, A. D., Jones, E. W. C. & Dawi, D. A. 1979 F_1 barley: male sterile multiplication and F_1 hybrid seed production. *Welsh Pl. Breed. Stn, A. Rep.* 1978, pp. 87–88.

Howard, H. W., Cole, C. S., Fuller, J. M., Jellis, G. J. & Thomson, A. J. 1977 Potato breeding problems with special reference to selecting progeny of cross Pentland Crown × Maris Piper. *A. Rep. Pl. Breed. Inst. Cambridge*, pp. 22–50.

Hughes, W. G. & Bodden, J. J. 1978 An assessment of the production and performance of F_1 hybrid wheats based on *Triticum timopheevi* cytoplasm. *Theor. appl. Genet.* **53**, 219–228.

Jinks, J. L. & Pooni, H. S. 1976 Predicting the properties of recombinant inbreds derived by single seed descent. *Heredity, Lond.* **36**, 253–266.

Kinman, M. L. & Sprague, G. F. 1945 Relation between number of parental lines and theoretical performance of synthetic varieties in corn. *J. Am. Soc. Agron.* **37**, 341.

Knott, D. R. & Kumar, J. 1975 Comparison of early generation yield testing and a single seed descent procedure in wheat breeding. *Crop Sci.* **15**, 295–299.

Lonquist, J. H. 1964 Modification of the ear-to-row procedure for the improvement of maize populations. *Crop Sci.* **4**, 227–228.

Lupton, F. G. H. & Whitehouse, R. N. H. 1957 Studies on the breeding of self-pollinating cereals. I. Selection methods in breeding for yield. *Euphytica* **6**, 169–184.

Ramage, R. T. 1965 Balanced tertiary trisomics for use in hybrid seed production. *Crop Sci.* **5**, 177–185.

Sage, G. C. M. 1976 Nucleo-cytoplasmic relationships in wheat. *Adv. Agron.* **28**, 265–298.

Simmonds, N. W. 1974 Costs and benefits of an agricultural research institute. *Res. Dev. Mgmt.* **5**, 23–28.

Sprague, G. F. & Eberhart, S. A. 1977 Corn breeding. In *Corn and corn improvement* (ed. G. F. Sprague), vol. 18, p. 714. American Society of Agronomy.

Sunderland, N. 1978 Comparative studies of anther and pollen culture. In *Plant and cell tissue culture: principles and applications* (ed. W. R. Sharp *et al.*), pp. 203–219. Ohio State University Press.

Thompson, K. F. 1964 Triple cross hybrid kale. *Euphytica* **13**, 173–177.

Wellhausen, E. J. 1978 Recent developments in maize breeding in the tropics. In *Maize breeding and genetics* (ed. D. B. Walden), p. 794. New York: Wiley.

Williams, W. & Gilbert, N. 1958 Heterosis on the inheritance of yield in the tomato. *Heredity, Lond.* **14**, 133–149.

Williams, W. 1959 Selection of parents and family size in the breeding of top fruits. In *Rept. 2nd Congr. Eucarpia*, pp. 221–223.

Discussion

D. R. Knott (*Crop Science Department, University of Saskatchewan, Saskatoon, Canada*). I wondered why Professor Williams indicated that lines developed by the single seed descent procedure should be carried to the F_8 or F_9. Adding several extra generations would reduce one of the main advantages of the single seed descent procedure, the reduction in the time required to produce a new cultivar.

W. Williams. One of the advantages of single seed descent is the achievement of homozygosity before selection, giving a better correspondence between **genotype** and **phenotype**. Selection in earlier generations could of course be practised, but this would reduce some of the value of single seed descent and such earlier selection would not give a timescale shorter than pedigree selection.

Physiological constraints to varietal improvement

By J. P. Cooper, F.R.S.

Welsh Plant Breeding Station, Plas Gogerddan, Aberystwyth, Dyfed SY23 3EB, U.K.

Crop production consists essentially of the conversion of environmental inputs into economic end-products, including human or animal foodstuffs or industrial raw materials. The basic climatic limitation to production is the seasonal input of solar energy, but the use of this energy by the crop can be limited by other climatic factors such as temperature or water supply, or by the supply of soil nutrients.

Both experimental results and predictive models indicate a potential fixation by a complete photosynthetic cover of 2–3 % of incoming energy into total biomass, corresponding to an annual dry matter production of up to 80 t/ha for tropical and 30 t/ha for temperate environments. Many crops, however, occupy the ground for only part of the year, and their production is influenced greatly by the duration of the photosynthetic canopy. Economic yield is further determined by the harvest index, i.e. the relative partition of assimilates to the economic end product.

Appreciable genetic variation has been revealed for many of the physiological components of crop photosynthesis and of the distribution and use of assimilates, including their response to temperature and water stress. Its effective use in a breeding programme depends on the identification of those components that are most important in determining yield or quality, and the development of rapid and reliable screening procedures that correlate well with the performance of the crop in the field.

1. Introduction

Crop production consists essentially of the conversion of environmental inputs such as light energy, CO_2, water and soil nutrients into economic end-products, which may be human or animal foodstuffs or industrial raw materials. The plant breeder is concerned to improve the efficiency of this process, i.e. to maximize the output from the available inputs in biological or economic terms (Wallace *et al.* 1972; Cooper 1975; Evans 1975). He therefore needs to ask (1) what the input limitations are, (2) what the current efficiency of the crop is in dealing with them, and (3) what characteristics of the crop contribute to this efficiency and how far they can be manipulated by the plant breeder.

The basic climatic limitation to crop production is the seasonal input of light energy, but the use of this energy can be limited by other climatic factors, particularly low temperature and shortage of water, by the availability of soil nutrients or by pest and disease attack. The energy input varies with latitude and with cloud cover, being comparatively uniform through the year in equatorial regions but showing increasing seasonal amplitude with increasing latitude. The highest inputs, over 700 kJ cm^{-2} per year are found in subtropical regions with little cloud cover, while most north temperate regions have comparatively low inputs, about 400 kJ cm^{-2}. per year, but with a large seasonal amplitude.

Even for a complete photosynthetic cover, however, the proportion of incoming radiation fixed into total biomass is comparatively small (Monteith 1972; 1977). Only some 50 % of the total radiation is photosynthetically active, while the quantum requirement for photosynthesis allows a maximum potential fixation of only about 10 % of the total input. In crops with the usual C_3 photosynthetic pathway (though not in C_4 plants), photorespiration results in a further loss of some 30–40%. Furthermore, at moderate to high light intensities, leaf and crop photosynthesis become CO_2-limited, with further reduction in energy fixation, while dark respiration usually reduces the *gross* photosynthesis by some 40–50%. Both experimental records (Cooper 1975; Loomis & Gerakis 1975) and crop models (de Wit 1965; Duncan *et al.* 1967; Monteith 1977) indicate that closed crop canopies can usually fix only some 2–3% of the total incoming energy, corresponding to maximum crop growth rates of 20–30 g m^{-2} day^{-1} for temperate C_3 species and over 40 g m^{-2} per day for C_4 species in the subtropics. If maintained over the year, such fixation could provide a total biomass of up to 80 t ha^{-1} per year in the subtropics and some 25–30 t ha^{-1} per year in more temperate regions, and such values have in fact been reported for perennial crops (Cooper 1975; Loomis & Gerakis 1975).

Most crops, however, do not provide a complete crop cover through the year, nor does the economic end-product usually consist of the total biomass. In the small grain cereals, for instance, the crop occupies the ground for only part of the year, and the dry matter in the grain is derived largely from current photosynthesis during or just before the period of grain filling.

The plant breeder, therefore, needs not only to provide an efficient photosynthetic cover but to maintain it for the optimum duration and to ensure the optimum partition of assimilates to the economic end-product, i.e. to maximize the harvest index.

2. Physiological components of production

In attempting to improve the physiological efficiency of his crop, the breeder needs to identify those features of the plant that contribute to such efficiency, to determine how much variation exists for them and to consider how far this variation can be used in a breeding programme.

(a) Crop photosynthesis

The photosynthetic activity of the crop will be influenced both by the photosynthetic rate of the individual leaves and by their arrangement in the crop canopy in relation to light interception.

(i) Individual leaf photosynthesis

The responses of the individual leaf to changes in light intensity, temperature and CO_2 supply are well documented (Wilson 1973; Troughton 1975; Cooper 1976). At low light intensities, photochemical processes are limiting and up to 10% of the total radiation may be fixed. As light intensity increases however, CO_2 transport or utilization become more important, and eventually at light saturation, photosynthesis becomes CO_2-limited and the maximum photosynthetic rate (P_{max}) is reached.

A major distinction in photosynthetic response is between species with the C_3 pathway (most temperate crops and such tropical crops as oil palm, rice and cassava) and those with the C_4 pathway (maize, sugar cane and most tropical forage grasses). Most C_3 species reach light saturation at about 100–150 J m^{-2} s^{-1} photosynthetically active radiation, less than half of full sunlight, with a P_{max} of about 50–100 ng CO_2 cm^{-2} s^{-1}. In C_4 species, on the other hand, the photosynthetic rate usually continues to increase up to inputs of over 300 J m^{-2} s^{-1}, i.e. approaching full sunlight, providing a P_{max} up to about 220 ng CO_2 cm^{-2} s^{-1}. This higher P_{max} in C_4 species is based in part on the greater affinity for CO_2 of the initial carboxylating enzyme (PEP carboxylase in C_4 plants and RuDP carboxylase in C_3), and in part on the absence of photorespiration, which in C_3 plants can amount to 30–40% of the *gross* photosynthesis (Hatch *et al.* (eds) 1971; Troughton 1975).

Marked varietal or genotypic differences in maximum photosynthetic rate per unit leaf area have been reported in many C_3 crops including soybean, rice, barley, *Phaseolus*, lucerne and ryegrass, as well as in such C_4 crops as maize, sugar cane and *Cenchrus ciliaris* (Wallace *et al.* 1972; Shibles *et al.* 1975; Wilson 1981). In few cases, however, was there any regular correlation with total or economic yield, suggesting that a lower photosynthetic rate per unit leaf area was compensated for by larger but thinner leaves (Evans & Dunstone 1970; Hart *et al.* 1978), or indeed that in these comparisons at least, dry matter production was not source-limited (Evans 1975).

Since much of the greater efficiency of the C_4 over the C_3 photosynthetic pathway can be attributed to its lack of photorespiration, it has been suggested that a decrease in the photorespiratory activity of C_3 species, possibly by uncoupling the carboxylase and oxygenase activity of RuDP carboxylase, could greatly increase their maximum photosynthetic rate (Zelitch 1976). In spite of extensive surveys, however, no individuals with C_4 characteristics have yet been found within the C_3 crop species, soybean, wheat and beet (Moss & Musgrave 1971), and crosses between C_3 and C_4 species of *Atriplex* indicate independent though complex genetic control of the various aspects of the C_4 syndrome (Björkman 1976). Quantitative variation in photorespiratory activities has, however, been recorded within a number of C_3 species, with in some cases favourable effects on net photosynthesis (Wilson 1972; Zelitch 1976).

(ii) *Leaf arrangement and canopy photosynthesis*

Most crops consist of a more or less closed canopy of leaves in which incoming light is distributed over a considerable leaf area, with a consequent increase in photosynthetic activity per unit area of ground compared with unit area of the individual leaf. In temperate C_3 crops, for instance, crop growth rates of about 20 g m^{-2} per day during May–September corresponding to a fixation of about 3–4% of the incoming photosynthetically active radiation are commonly attained, while similar crop growth rates and energy fixations have been reported for tropical C_3 species. For C_4 species, however, growing under conditions of high insolation, higher values of about 50 g m^{-2} per day, corresponding to over 6% fixation of photosynthetically active radiation have been recorded for such crops as maize, sudan grass and bulrush millet (Cooper 1975; Loomis & Gerakis 1975).

The penetration of light down the crop, the area of leaf that can be illuminated, and hence the potential crop photosynthesis, will be influenced by the arrangement of leaves in the canopy. An erect leaf arrangement, as in the cereals, for instance, allows the incoming light energy to be distributed over a greater leaf area than in a crop such as white clover or cotton with broad flat leaves (Saeki 1975; Monteith 1969).

Variation in leaf arrangement and canopy structure has been reported within many crops including maize, rice, soybean and forage grasses (Wallace *et al.* 1972), and in some cases varieties with more erect leaves have been found to possess higher crop growth rates or photosynthetic rates (Sheehy & Cooper 1973; Rhodes 1975; Austin *et al.* 1976). The optimum leaf arrangement will, however, depend on the stage of growth of the crop. During early establishment, a more prostrate habit of growth is advantageous to enable the interception of the maximum amount of light, but once complete interception has been achieved, an erect canopy, which spreads the incoming light energy over a larger leaf area, is likely to be more valuable.

(iii) *Respiratory losses*

A large part of the assimilates produced by photosynthesis, in many crops up to some 50%, are utilized in respiration (Robson 1973; Biscoe *et al.* 1975). Much respiration is likely to be coupled to active growth processes, but its separation into a 'biosynthetic' component, used for growth, and a 'maintenance' component has recently been posulated (McCree 1970; Penning de Vries 1974; Thornley 1977), raising the possibility of reducing the requirements for maintenance respiration. In the early stages of crop growth, much of the energy expenditure is concerned with active cell division and expansion, but as the crop develops the proportion of non-growing tissue increases, and 'maintenance' respiration becomes proportionately greater. Reduction in 'maintenance' respiration should therefore have its greatest influence at later states of crop growth, when there is a larger standing biomass. In fact, a negative relation between dark respiration of mature tissues and crop growth has been reported in a number of species including sorghum, ryegrass and barley (Heichel 1971; Wilson 1975b). Ryegrass, for instance, showed a significant response to selection for slow dark respiration of mature leaves with little effect on the growth of young seedlings, but considerably increased dry matter production in older plants (Wilson 1975b, 1976).

Even so, although appreciable genetic variation in many of the components of crop photosynthesis has been reported, in only a few cases has a consistent relation with total or economic yield been recorded. In practice, the duration of the photosynthetic canopy and the relative partition of assimilates to the economic yield, i.e. the harvest index, appear to be more important.

(b) *Leaf area duration*

The importance of leaf area duration in determining both total and economic yield was early demonstrated by Watson (1947) and, more recently, Monteith (1977) has shown a close relationship for a range of crops in the U.K. between total dry matter production and radiation intercepted by the canopy.

The possible duration of the growing season is often limited by local climatic factors, particularly temperature or water limitations and/or by the requirements of particular farming systems. Where sowing or planting are limited by low temperature, selection for a lower temperature threshold for germination or leaf growth may be an important breeding objective. In temperate crops such as potatoes it should allow of earlier planting and hence more effective use of increasing light energy in the spring (Moorby & Milthorpe 1975; Wareing & Allen 1977), while in subtropical crops such as maize or *Phaseolus* it may make possible their adaptation to a wider climatic range. Again, in temperate regions, selection for improved cold tolerance may make possible autumn rather than spring sowing, thereby providing a closed photosynthetic cover to take advantage of the increasing energy inputs in the spring. As mentioned later, considerable genetic variation exists in most crops for those temperature responses that influence germination, leaf growth and plant survival, and so determine sowing or planting dates.

Increasing leaf area duration by a later harvest date, when this does not conflict with other farming requirements, may also be advantageous, particularly in crops such as sugar beet and potato (Moorby & Milthorpe 1975), which can continue to transfer assimilates into the re-

It must not be forgotten, however, that variation in canopy structure may also have important effects on the distribution of CO_2, temperature and water vapour through the crop.

quired vegetative sinks. In determinate crops such as the cereals, on the other hand, the timing of flowering and seed production, and hence the termination of the active growing season, is usually based on responses to photoperiod and low-temperature vernalization (Evans et al. 1975). These responses can often be manipulated by the breeder, as in the adaptation of such short-day species as maize (Duncan 1975) and soybean (Shibles et al. 1975) to higher latitudes, or the wider regional adaptation of the day-neutral CIMMYT wheats (Evans 1975).

(c) *Partition of assimilates (harvest index)*

The most important determinant of economic yield is often not crop photosynthesis, i.e. the supply of assimilates, but the way in which they are distributed within the plant either for continued vegetative growth or for accumulation in particular sinks such as storage organs, seeds or fruits. It is often not clear how far economic yield is limited by the supply of assimilates i.e. source strength, by the ability of the sinks to make use of them, or by the rate of translocation, nor indeed how far sink strength can itself influence photosynthetic rates (Evans 1975; Wareing & Patrick 1975).

Apart from those crops in which the leaf canopy itself forms the economic yield, as in green vegetables and forage grasses, an important distinction is between crops with a strictly determinate habit, as in the cereals, which show a clear separation of vegetative and reproductive allocation of assimilates, and those such as potatoes, sugar beet and sugar cane, which can continue indefinite growth and allocation of assimilates to a vegetative sink. In yet other crops such as cotton and field beans, with an indeterminate habit, allocation to vegetative growth and to flowering and seed production can continue simultaneously (Evans 1975).

In crops with vegetative sinks, a major objective is the prolongation of the growing season together with a maximum partition of assimilates to the required end-product, since both total and economic yield are likely to be proportional to the total energy intercepted. In the potato, for example, marked varietal differences exist in both the onset and rate of bulking (Moorby & Milthorpe 1975). In determinate crops, on the other hand, the appropriate balance of source and sink is more complex. In wheat and barley, the content of the grain results largely from current assimilation by the ear and flag leaf, with some transfer from the stem, but the potential sink size, i.e. the number of grain initials, is determined during ear differentiation (Evans et al. 1975; Austin & Jones 1976; Biscoe et al. 1975). Maize, on the other hand, shows a considerable transfer of assimilates from stem to grain, and crop photosynthesis can contribute directly to grain yield over a longer period (Adelana & Milbourne 1972; Duncan 1975). In the indeterminate group of crops, such as cotton, soybean and field bean, competition between continued vegetative growth and reproductive sinks often results in a rather unpredictable yield from year to year; the introduction of the determinate habit, as is now possible in field beans, should improve harvest index and reliability of yield (Shibles et al. 1975).

Considerable variation in harvest index is apparent in most crops, and selection for such features as onset and rate of tuberization in potatoes (Moorby & Milthorpe 1975), high root or sugar yield in beet, and short straw or restricted tillering in cereals (Donald & Hamblin 1976) continue to be important breeding objectives. Even so, comparatively little is known of the physiological basis of those processes that determine harvest index, including the translocation of assimilates, their partition to different morphological sinks in the plant, or, in certain crops, to particular biochemical end-products.

(d) Response to climatic stress

Both the duration of the growing season and the rate of photosynthesis and growth during that season can be limited by temperature or water shortage. Furthermore, the breeder is often presented with two conflicting requirements, continued active growth at moderate stress, and/or survival of more extreme conditions, often involving a degree of dormancy.

(i) Temperature

The temperature responses for both photosynthesis and leaf growth are usually related to the climatic origin of the crop, most temperate species having a broad temperature optimum between 15 and 25 °C though some photosynthesis and extension growth can occur at temperatures as low as 5 °C. Net photosynthesis and active extension growth are usually greatly reduced at temperatures above 30–35 °C. Most tropical and subtropical species (whether C_3 or C_4), on the other hand, have higher temperature optima, up to 35 °C or more, and both photosynthesis and leaf extension growth are usually limited at temperatures below 15 °C (Bauer et al. 1975; Cooper 1976).

Even within the same species, however, populations from different climatic regions may differ in their temperature response, offering scope for the plant breeder. In temperate forage grasses, for instance, populations from the Mediterranean region, where winter is the active growing season, can expand leaves actively at moderately low temperatures (0–5 °C), while collections from northern or central Europe show a degree of winter dormancy, associated with increased winter hardiness (Cooper 1964; Østgård & Eagles 1971). In annual temperate crops, more active leaf growth at lower temperatures can be valuable in improving light interception in the spring, while in such subtropical crops as maize and *Phaseolus*, variation in temperature response is important in increasing the climatic range of these crops.

The survival of more extreme cold can be achieved by two contrasting strategies. First, the life cycle may be timed to avoid the period of extreme stress, as in many annual crops in north temperate or continental climates. The spring cereals, for instance, are spring-sown and harvested in late summer or early autumn, while potatoes or sugar beet, though potentially biennial or perennial, are grown as spring-planted annuals. Secondly, the crop may develop sufficient cold hardiness, often associated with dormancy to survive the unfavourable season. Many temperate tree crops cease extension growth in the winter, and similar seasonal dormancy patterns are shown by forage crops or winter cereals adapted to higher latitudes. The timing of such dormancy in locally adapted varieties is usually related to the length and severity of the winter, and is often influenced by response to other climatic factors such as photoperiod, thus enabling the plant to become dormant before conditions become too severe (Wareing 1969; Østgård & Eagles 1971; Bauer et al. 1975).

(ii) Water stress

Water stress can be induced by shortage of water supply to the roots and/or by excessive water demand from the leaves, and often coincides with periods of high insolation and hence high potential crop photosynthesis. In environments with moderate or intermittent water shortage, continued growth may involve the ability to tap greater reserves of soil water by increase in root range, as in bulrush millet compared with sorghum (Wetselaar & Norman 1960), and/or the conservation of water by the control of transpiration. An important advantage of the

C_4 photosynthetic pathway, for instance, is increased efficiency in water use, based both on a higher photosynthetic rate and a greater stomatal resistance to water vapour transfer (Downton 1971), while a more extreme conservation strategy is shown by species with the Crassulacean acid metabolism photosynthetic system, such as pineapple, in which the stomata can close during the high evaporative demand of the day (Troughton 1975).

Even within a single species, however, variation has been reported for certain leaf characteristics that influence transpiration and water use efficiency. In barley (Misken *et al.* 1972; Yoshida 1978) and *Panicum antidotale* (Dobrenz *et al.* 1969), a lower stomatal frequency can lead to reduced transpiration without any corresponding reduction in photosynthesis, while in perennial ryegrass lines selected for smaller stomata or fewer stomata per unit surface area can continue active leaf growth for a longer period when exposed to moderate water stress (Wilson 1975a, 1981).

In more extreme environments with long and regular periods of drought, the two contrasting strategies of avoidance and/or resistance have been developed. In a Mediterranean environment, for instance, many grain and forage crops germinate with the autumn rains, grow actively through the winter, and flower and produce seed in the spring before the summer drought stops further growth. In these crops, the temperature and photoperiod requirements for flowering and seed production of locally adapted varieties are closely tailored to the length of the potential growing season (Cooper & McWilliam 1966; Finlay & Wilkinson 1963). On the other hand, perennial grasses such as *Phalaris tuberosa* and *Hordeum bulbosum* become dormant during the Mediterranean summer while maintaining a deep rooting system (McWilliam 1968; Koller 1969). Even so, in spite of its great agricultural significance, comparatively little is known of the physiological basis of variation in response to water stress.

In conclusion, appreciable genetic variation appears to exist for many of the physiological features that affect crop photosynthesis, leaf area duration, harvest index and response of the plant to climatic stresses. How far can this variation be consciously used in a breeding programme?

3. Use of physiological variation in a breeding programme

In practice, the plant breeder is concerned with the performance of his crop in the field, and any physiological screening procedures must not only provide more effective assessment than can be obtained in the field, but also show a good correlation with yield or quality. He therefore needs to consider not only which particular physiological features are important in determining crop yield but how far they can be screened rapidly and reliably, and what advantages such screening has over conventional field or glasshouse assessment (Wallace *et al.* 1972; Cooper 1974).

The relative importance of any physiological feature can be assessed in three complementary ways.

1. The analysis of contrasting climatic or agronomic varieties or collections. Much of our knowledge of the physiological basis of response to climatic stress, of developmental responses to temperature and photoperiod, of leaf area duration and of the distribution of assimilates, i.e. harvest index, has been gained from such comparative studies.

2. Selection from within a variety or segregating population for high and low expression of a particular physiological feature. The direct effect of such selection on performance in the field

can then be assessed, while any correlated responses, either favourable or unfavourable, will become apparent. Furthermore, those selection lines that show promise can be readily incorporated into a breeding programme.

3. The use of models of crop photosynthesis and growth to predict the effects on yield of changes in such features such as individual leaf photosynthesis, canopy structure and the distribution of assimilates. Such models are rapidly being developed, though as yet they have hardly attained the degree of precision required for use in a breeding programme (de Wit 1965; Monteith 1972, 1977).

Even where the importance of a particular physiological feature has been demonstrated, however, any operational screening tests must be able to handle large numbers of individuals rapidly and reliably, preferably early in the life cycle and at any time of the year, and, of course, show a good correlation with subsequent performance in the field. A reasonable heritability with no unfavourable correlated responses is also important.

At present, early screening tests are likely to be particularly valuable for stress response, where field assessment is restricted to one season of the year and can be influenced by unpredictable environmental fluctuations. In many crops, particularly the cereals and forage grasses, seedling tests under controlled conditions have been developed to assess either cold hardiness or leaf growth at moderately low temperatures, which relate well to subsequent performance in the field. Direct methods of screening for response to water stress are less well developed, but indirect techniques such as selection for smaller stomata or fewer stomata per unit area in the forage grasses have led to improved water use efficiency and a marked increase in summer production (Wilson 1975a, 1981).

Similarly, those developmental responses to low-temperature vernalization or photoperiod that control the onset of flowering or tuberization can often be measured reliably under controlled conditions, though since the seasonal cycle of photoperiod varies regularly in the field, any advantage over field assessment for a single location is more doubtful. Even so, such screening may be of value in predicting the flowering behaviour of the material in other environments, and can lead to the development of controlled techniques for the acceleration of flowering and hence a more rapid turnover of generations.

The potential value of screening for components of crop photosynthesis is rather less clear (Wallace et al. 1972). Although variation in the maximum photosynthetic rate of the individual leaf has been recorded in many crops, in few cases has it been possible to relate this to differences in total dry matter or in economic yield. The effect of variation in photorespiration in C_3 crop species is also not clear, though on theoretical grounds a possible reduction of photorespiration could greatly increase dry matter production and crop yield. Furthermore, rapid and reliable laboratory techniques for the routine screening of large numbers of individuals for photosynthesis or photorespiration still remain to be developed.

More information, however, is accumulating on the possible effects of reducing maintenance respiration. In the forage grasses, for instance, selection for reduced dark respiration of mature leaves has resulted in an increase of some 12% in annual dry matter production, the increase being particularly marked in the higher temperatures of summer (Wilson 1975b, 1976).

Similarly, considerable variation has been recorded within crop species for such canopy characters as leaf angle, leaf length and leaf rigidity, which determine the distribution of light down the canopy, but, as discussed earlier, the effects of such differences vary with the crop and the stage of growth. Even so, in the forage grasses, selection for such characters in the early seedling stage has resulted in marked changes in light distribution and in maximum crop growth rate (Rhodes 1975).

In general, however, the extent to which variation in total biomass, and certainly in economic yield, is determined by variation in net photosynthesis of the crop canopy is far from clear. In practice, differences in the distribution and use of assimilates either to provide more effective light interception, i.e. a greater leaf area duration, or more effective partition to economic yield, i.e. a greater harvest index, would appear to be more important (Evans 1975). It is in these two areas that comparative physiological studies are likely to lead to greater understanding and improvement of crop yield.

Furthermore, the present discussions have been concerned primarily with the limitations to crop improvement set by the main climatic variables of energy input, temperature and water supply. Looking ahead, the cost and availability of other inputs, including the major nutrients nitrogen and phosphorus, may become important limitations, and the physiological efficiency of the crop in dealing with these inputs also requires serious study.

References (Cooper)

Adelana, B. O. & Milbourne, G. M. 1972 The growth of maize. II. Dry matter partition in three maize hybrids. *J. agric. Sci., Camb.* **78**, 73–78.

Austin, R. B. & Jones, R. B. 1976 The physiology of wheat. *Cambridge Pl. Breed. Inst. Rep. 1975*, pp. 20–73.

Austin, R. B., Ford, M. A., Edrich, J. A. & Hooper, B. E. 1976 Some effects of leaf posture on photosynthesis and yield in wheat. *Ann. appl. Biol.* **83**, 425–446.

Bauer, H., Larcher, W. & Walker, R. B. 1975 Influence of temperature stress on CO_2 gas exchange. In *Photosynthesis and productivity in different environments* (ed. J. P. Cooper), pp. 557–586. Cambridge University Press.

Biscoe, P. V., Scott, R. K. & Monteith, J. L. 1975 Barley and its environment. III. Carbon budget of the stand. *J. appl. Ecol.* **12**, 269–293.

Björkman, O. 1976 Adaptive and genetic aspects of C_4 photosynthesis. In *CO_2 metabolism and plant productivity*. (ed. R. H. Burris & C. C. Black), pp. 287–310. Baltimore: University Park Press.

Cooper, J. P. 1964 Climatic variation in forage grasses. I. Leaf development in climatic races of *Lolium* and *Dactylis*. *J. appl. Ecol.* **1**, 45–61.

Cooper, J. P. 1975 Control of photosynthetic production in different environments. In *Photosynthesis and productivity in different environments* (ed. J. P. Cooper), pp. 593–612. Cambridge University Press.

Cooper, J. P. 1976 Photosynthetic efficiency of the whole plant. In *Food production and consumption: the efficiency of human food chains and nutrient cycles* (ed. A. N. Duckham, J. G. W. Jones & E. H. Roberts), pp. 107–126. Amsterdam: North-Holland.

Cooper, J. P. & McWilliam, J. R. 1966 Climatic variation in forage grasses. II. Germination, flowering and leaf development in Mediterranean populations of *Phalaris tuberosa*. *J. appl. Ecol.* **3**, 191–212.

Dobrenz, A. K., Wright, L. N., Humphrey, A. B., Massengale, M. A. & Kneebone, W. R. 1969 Stomate density and its relationship to water-use efficiency of blue panicgrass (*Panicum antidotale* Retz.). *Crop Sci.* **9**, 354–357.

Donald, C. M. & Hamblin, J. 1976 The biological yield and harvest index of cereals as agronomic and plant breeding criteria. *Adv. Agron.* **28**, 361–405.

Downton, W. J. S. 1971 Adaptive and evolutionary aspects of C_4 photosynthesis. In *Photosynthesis and photorespiration* (ed. M. D. Hatch, C. B. Osmond & R. O. Slatyer), pp. 3–17. New York: Wiley–Interscience.

Duncan, W. G. 1975 Maize. In *Crop physiology: some case histories* (ed. L. T. Evans), pp. 23–50. Cambridge University Press.

Duncan, W. G., Loomis, R. S., Williams, W. A. & Hanau, R. 1967 A model for simulating photosynthesis in plant communities. *Hilgardia* **38**, 181–205.

Evans, L. T. 1975 The physiological basis of crop yield. In *Crop physiology: some case histories* (ed. L. T. Evans), pp. 327–355. Cambridge University Press.

Evans, L. T. & Dunstone, R. L. 1970 Some physiological aspects of evolution in wheat. *Aust. J. biol. Sci.* **23**, 725–741.

Evans, L. T., Wardlaw, I. F. & Fischer, R. A. 1975 Wheat. In *Crop physiology: some case histories* (ed. L. T. Evans), pp. 101–149. Cambridge University Press.

Finlay, K. W. & Wilkinson, G. N. 1963 An analysis of adaptation in a plant breeding programme. *Aust. J. agric. Res.* **14**, 742–754.

Hart, R. H., Pearce, R. B., Chatterton, N. J., Carlson, G. E., Barnes, D. K. & Hanson, C. H. 1978 Alfalfa yield, specific leaf weight, CO_2 exchange rate and morphology. *Crop. Sci.* **18**, 649–653.

Hatch, M. D., Osmond, C. B. & Slatyer, R. O. (eds) 1971 *Photosynthesis and photorespiration.* New York: Wiley-Interscience.

Heichel, G. H. 1971 Genetic control of epidermal cell and stomatal frequency in maize. *Crop Sci.* **11**, 830–832.

Koller, D. 1969 The physiology of dormancy and survival of plants in desert environments. *Symp. Soc. exp. Biol.* **23**, 449–470.

Loomis, R. S. & Gerakis, P. A. 1975 Productivity of agricultural ecosystems. In *Photosynthesis and productivity in different environments* (ed. J. P. Cooper), pp. 145–172. Cambridge University Press.

McCree, K. J. 1970 An equation for the rate of respiration of white clover plants grown under controlled conditions. In *Prediction and measurement of photosynthetic productivity* (ed. I. Setlik), pp. 221–230. Wageningen: Centre for Agricultural Publishing and Documentation.

McWilliam, J. R. 1968 The nature of the perennial response in Mediterranean grasses. II. Senescence, summer dormancy and survival in *Phalaris*. *Aust. J. agric. Res.* **19**, 397–410.

Miskin, K. E., Rasmusson, D. C. & Moss, D. N. 1972 Inheritance and physiological effects of stomatal frequency in barley. *Crop Sci.* **12**, 780–783.

Monteith, J. L. 1969 Light interception and radiative exchange in crop stands. In *Physiological aspects of crop yield* (ed J. D. Eastin, F. A. Haskins, C. Y. Sullivan & C. H. M. van Bavel), pp. 89–111. Madison, Wisconsin: A.S.A. and C.S.S.A.

Monteith, J. L. 1972 Solar radiation and productivity in tropical ecosystems. *J. appl. Ecol.* **9**, 747–766.

Monteith, J. L. 1977 Climate and the efficiency of crop production in Britain. *Phil. Trans. R. Soc. Lond.* B**281**, 277–294.

Moorby, J. & Milthorpe, F. L. 1975 Potato. In *Crop physiology: some case histories* (ed. L. T. Evans), pp. 225–257. Cambridge University Press.

Moss, D. N. & Musgrave, R. B. 1971 Photosynthesis and crop production. *Adv. Agron.* **23**, 317–336.

Østgård, O. & Eagles, C. F. 1971 Variation in growth and development of natural populations of *Dactylis glomerata* from Norway and Portugal. II. Leaf development and tillering. *J. appl. Ecol.* **8**, 383–392.

Penning de Vries, F. W. T. 1974 Substrate utilization and respiration in relation to growth and maintenance in higher plants. *Neth. J. agric. Sci.* **22**, 40–44.

Rhodes, I. 1975 The relationships between productivity and some components of canopy structure in ryegrass (*Lolium* spp.). IV. Canopy characters and their relationship with sward yields in some intra-population selections. *J. agric. Sci., Camb.* **84**, 345–351.

Robson, M. 1973 The growth and development of simulated swards of perennial ryegrass. II. Carbon assimilation and respiration in a seedling sward. *Ann. Bot.* **37**, 501–518.

Saeki, T. 1975 Distribution of radiant energy and CO_2 in terrestrial communities. In *Photosynthesis and productivity in different environments* (ed. J. P. Cooper), pp. 297–322. Cambridge University Press.

Sheehy, J. E. & Cooper, J. P. 1973 Light interception, photosynthetic activity, and crop growth rate in canopies of six temperate forage grasses. *J. appl. Ecol.* **10**, 239–250.

Shibles, R. M., Anderson, I. C. & Gibson, A. H. 1975 Soybean. In *Crop physiology: some case histories* (ed. L. T. Evans), pp. 151–189. Cambridge University Press.

Thornley, J. H. M. 1977 Growth, maintenance and respiration: a re-interpretation. *Ann. Bot.* **41**, 1191–1203.

Troughton, J. H. 1975 Photosynthetic mechanisms in higher plants. In *Photosynthesis and productivity in different environments* (ed. J. P. Cooper), pp. 357–391. Cambridge University Press.

Wallace, D. H., Ozbun, J. L. & Munger, H. M. 1972 Physiological genetics of crop yield. *Adv. Agron.* **24**, 97–146.

Wareing, P. F. 1969 The control of bud dormancy in seed plants. *Symp. Soc. exp. Biol.* **23**, 241–262.

Wareing, P. F. & Allen, E. J. 1977 Physiological aspects of crop choice. *Phil. Trans. R. Soc. Lond.* B **281**, 107–119.

Wareing, P. F. & Patrick, J. 1975 Source–sink relations and the partition of assimilates in the plant. In *Photosynthesis and productivity in different environments* (ed. J. P. Cooper), pp. 481–500. Cambridge University Press.

Watson, D. J. 1947 Comparative physiological studies on the growth of field crops. *Ann. Bot.* **11**, 41–76.

Wetselaar, R. & Norman, M. J. T. 1960 Recovery of available soil nitrogen by annual fodder crops at Katherine, Northern Territory. *Aust. J. agric. Res.* **11**, 593–704.

Wilson, D. 1972 Variation in photorespiration in *Lolium*. *J. exp. Bot.* **23**, 517–524.

Wilson, D. 1973 Physiology of light utilization by swards. In *Chemistry and biochemistry of herbage* (ed. G. W. Butler & R. W. Bailey), vol. 2, pp. 57–101. London: Academic Press.

Wilson, D. 1975a Leaf growth, stomatal diffusion resistances and photosynthesis during droughting of *Lolium perenne* populations selected for contrasting stomatal length and frequency. *Ann. appl. Biol.* **79**, 67–82.

Wilson, D. 1975b Variation in leaf respiration in relation to growth and photosynthesis of *Lolium*. *Ann. appl. Biol.* **80**, 323–328.

Wilson, D. 1976 Physiological and morphological selection criteria in grasses. In *Breeding methods and variety testing in forage plants* (ed B. Dennis) pp. 9–18. Proc. Fodder Crops Section, Eucarpia, Roskilde, Denmark.

Wilson, D. 1981 Breeding for morphological and physiological traits. In *Plant Breeding Symposium* II (ed. K. J. Frey). Iowa State University Press. (In the press.)

de Wit, C. T. 1965 Photosynthesis of leaf canopies. *Versl. Landbouwk. Onderz. Ned.* **663**, 1–57.

Yoshida, T. 1978 On the stomatal frequency in barley. V. The effect of stomatal size on transpiration and photosynthetic rate. *Jap. J. Breed.* **28** (2), 87–96.

Zelitch, I. 1976 Biochemical and genetic control of photorespiration. In CO_2 *metabolism and plant productivity* (ed. R. H. Burris & C. C. Black), pp. 343–358. Baltimore: University Park Press.

The achievements of conventional plant breeding

By J. Bingham, F.R.S.

Plant Breeding Institute, Maris Lane, Trumpington, Cambridge CB2 2LQ, U.K.

Varietal improvements in yield are often strongly associated with increases in the ratio of the harvested organ or product to total biomass, which itself has shown relatively little change. Thus the greatest opportunities for increasing yield have been for the cereal species, with root crops intermediate and forages the most difficult to improve. In favourable climates the yields of cereals are frequently double those obtained in 1950, improvements in varieties accounting for about half this increase. The work has been much less effective where growth is dominated by environmental stresses, especially water supply, and in these situations yields are inconsistent, with little average improvement.

Breeding for resistance to the hazards of pests and disease, storms, temperature extremes and mineral deficiencies has had numerous successes. There remain considerable problems in the durability of resistance to pathogens, and in a number of cases there are no known sources of resistance that can be utilized by conventional breeding methods.

Improvements in the quality of food crops are most notable for meeting technological requirements and consumer preferences; there are few examples of improvement in nutritional value other than in the elimination of nutritionally deleterious substances. In cereals and many other crops there are considerable limitations in protein content due to an inverse relation with yielding ability.

Introduction

Plant breeding has contributed significantly to all the major crops of agriculture and horticulture, in many cases enabling the grower to obtain higher yields, adopt more efficient methods of production and meet the demands of the processor or consumer for increasingly close specifications in the harvested product. In the context of this meeting, little would be gained by attempting to catalogue these achievements. It is more appropriate to consider them with the particular purpose of identifying areas of plant breeding in which the present work may be considered wanting.

First, there are many examples of characters in which varietal improvements have made a major contribution and may be expected to continue to do so. In such cases new methods must have a cost effective advantage in rate of progress or in more efficient use of the breeder's time and facilities. The new methods will have to compete with well established practices, which often amount to massive experiments in genetics, though the application of the work may owe little to detailed genetic information. For example, in cereals it is now common for a breeding group to handle all stages of a programme based on about 1 000 000 plants in F_2 and 40 000 single plant progenies in F_3 (Bingham 1979).

Secondly, it may be possible to delineate some objectives with serious limitations in prospect for improvement by conventional methods, either in the rate of progress or in the ultimate level of expression of a desirable character. Breeding problems in this area may be due to a lack of genetic resources, to cytogenetic barriers in their use or to the lack of effective selection methods. Characters where the rate of progress is now decelerating after a period of rapid advance should be included in this group.

In following this general theme, examples will be drawn mostly from those agricultural and horticultural crops grown as food for man and animals. The objectives will be considered in the interrelated classes of yield potential, resistance to environmental hazards, and quality of the product.

VARIETAL IMPROVEMENT IN YIELD POTENTIAL

The average yields obtained by growers of many agricultural and horticultural crops have doubled over the last 30 years. For wheat (figure 1), yields in the United Kingdom increased slowly over the first 50 years of this century from about 2.1 to 3.0 t/ha; they have since advanced more rapidly to an estimated 5.7 t/ha (at 15% moisture) in 1980, with below-average years due to bad harvest weather in 1968 and to drought in 1976.

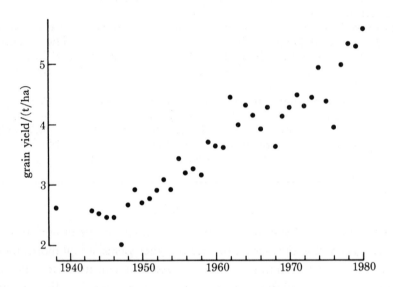

FIGURE 1. National average yields (15% moisture content) for wheat in the United Kingdom.

The highest recorded farm yields in the United Kingdom have increased correspondingly. For example, in 1980, R. Harvey at Braiseworth Hall Farms, Tannington, Suffolk, obtained 11.5 t/ha (15% moisture) for a 83 ha field of the winter wheat variety Hustler, on a deep, moisture retentive soil of the Beccles Series, with an average of 10.9 t/ha over 420 ha. J. Muirhead, on a deep alluvial soil at Eastlands Farm, Bradwell on Sea, Essex, also recorded 11.5 t/ha for 10.4 ha of the variety Virtue. These yields are close to the theoretical maximum for present-day varieties in this country of 11.4–12.9 t/ha, calculated by Austin (1978) on the basis of measurements of photosynthesis and the proportion of assimilates used in grain filling.

The increases in yield have been obtained over a period of rapidly changing husbandry practices, with greatly increased inputs of fertilizers, herbicides and fungicides. For the self-fertilized cereal species, the proportion of the increase in yield due to improvements in varieties may be assessed by direct comparison of varieties. For the period 1947–75, Silvey (1978) estimated, from the yields of controls in National Institute of Agricultural Botany (N.I.A.B.)

trials, that varietal improvement accounted for increases of 50% in national average yields for wheat and 30% for barley. These calculations compound the effects of potential yielding ability with resistance to lodging and disease resistance, and may be an overestimate because standards of comparison change owing to increases in the virulence of pathogens on varieties used as controls.

To isolate the effects of varietal improvement in yield potential from resistance to hazards, Austin *et al.* (1980) compared winter wheat varieties, representing a chronological series, in yield trials protected by fungicides and supported by nets to prevent lodging. The trials were grown with high and low levels of nitrogen fertilizer application. Apart from one line, Benoist 10483, the yields for the two levels of soil fertility were strongly correlated (figure 2), indicating

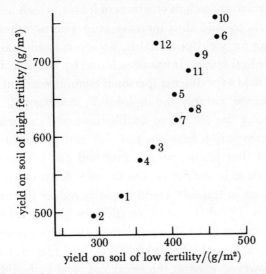

FIGURE 2. Grain yield (dry mass) of a series of winter wheat varieties when grown on soil of high fertility, receiving 104 kg N/ha, and low fertility, receiving 38 kg N/ha (Austin *et al.* 1980). Varieties and year of introduction: 1, Little Joss (1908); 2, Holdfast (1935); 3, Cappelle-Desprez (1953); 4, Maris Widgeon (1964); 5, Maris Huntsman (1972); 6, Hobbit (1977); 7, Mardler (1978); 8, 370/500(†); 9, Galahad(†); 10, Norman (1981); 11, Armada (1978); 12, Benoist 10483(†). †, Breeder's lines or varieties not grown commercially.

that percentage advantage in yield of the newer varieties was independent of nitrogen application, Hobbit and Norman outyielding the average of Little Joss and Holdfast by 45–49% in each treatment. The higher yielding varieties were, however, more responsive and efficient users of nitrogen in the sense that they gave a greater return in grain yield per kilogram N applied. This effect on yield potential is additional to the practicality of using higher rates of nitrogen fertilizer on varieties with improved standing ability.

The varietal improvement in grain yield in this series of varieties was associated predominantly with increases in harvest index (ratio of grain yield to that of straw and grain combined). There were only small differences between varieties in total dry matter production, but these were not related to grain yield, so that there was a strong tendency for higher yielding varieties to have lower straw masses per square metre. The varieties were similar in maximum leaf area index, and yielding ability was not associated with any particular combination of the harvest components of ear number, ear size or 1000 grain mass. T. J. Riggs (personal communication) obtained similar results with a series of spring barley varieties in experiments carried out in

1979 and 1980. Other investigations have also indicated that change in dry matter distribution is the most important difference between old and modern varieties.

In recent years, increases in the harvest index of wheat varieties have depended mainly on the major semi-dwarfing genes *Rht*1 and *Rht*2 derived from Norin 10, and on other factors from Italian varieties. Such factors can be combined to give further reductions in height, but there remains only limited scope for increasing yield by substituting grain for straw. This approach is likely to become counter-productive if much shorter lines are developed, owing to adverse effects on the spatial arrangement of the leaves, and in taller lines because diversion of structural materials from the straw would reduce resistance to lodging.

More attention should therefore be given to increasing total biomass. It is probable that some increase in biomass can be achieved by selecting within the available variation in *Triticum aestivum* for morphological characters, such as erectness of leaves, which improve the uniformity of light distribution in the canopy and thus increase crop photosynthesis. The greatest remaining challenge in breeding for yield potential is, however, to increase the light-saturated rate of photosynthesis of individual leaves. It has been found by Evans & Dunstone (1970), and confirmed by R. B. Austin in field experiments (personal communication), that some primitive species of *Aegilops* and of *Triticum*, notably the diploids *T. thaoudar* and *T. urartu*, have rates of photosynthesis per unit area of flag leaf up to double those of *T. aestivum* varieties. There is, however, a strong negative correlation between leaf size and photosynthetic rate. Evans & Dunstone therefore concluded that photosynthetic rate had not yet limited the evolution of yield in wheat but might well do so in the future. For the very dense crops now grown on fertile soils it is unlikely that reductions in leaf size would seriously reduce light interception. As such crops become more common it is probable that selection for photosynthetic rate will become of increasing importance. It has yet to be shown whether the high photosynthetic rate of the diploid species can be maintained at the hexaploid level, and this may not be so if it is dependent on small cell size. However, even at the hexaploid level it should be possible to make some progress by deliberate selection for photosynthetic rate.

Taking these physiological considerations into account, there is no reason to suppose that a ceiling has been reached in yielding ability with present-day varieties of wheat and barley. This view is supported by genetical studies; for example, Laabassi (1979), using a recessive nuclear gene to produce F_1 wheat hybrids, found that heterosis for yield was commonly 15–20 % above the higher-yielding parent, indicating that it should be possible to obtain increases of at least this order by pure line breeding. Nevertheless, as genetic variation at the varietal level is exploited, a diminishing return in potential yield is inevitable and it is significant that maintenance of the present rate of progress is involving larger breeding teams and intensification of selection methods.

Increases in the on-farm yields of other crops in many cases equal those for cereals, but it is often more difficult to estimate the contribution of varietal improvement. Varieties of outbreeding species sown from seed may have been reselected or reconstituted during their period of use, and even comparatively recent varieties may no longer exist in their original form. In maize, many of the inbred lines used in the production of double and single cross hybrids have been maintained in collections, so it is possible to remake varieties grown since hybrids were first widely grown in the late 1920s. Russell (1974), working in Iowa, found that optimum seeding rates were higher for new than for old varieties of maize because they produced fewer barren stems. When sets of varieties representing different periods of cultivation were grown at a range of seed rates, those of the 1970s outyielded those of the 1930s by 38% at the optimum seed rate for each group and by 57% when averaged over all sowing rates. Thus the genetic gain in yield for maize in Iowa has been similar to that for wheat and barley in the United Kingdom.

When other crop species are compared with cereals, it is evident that those already giving the highest yields of harvested product are the most difficult to improve further, especially when the high yields are associated with high harvest index. Rating of crops grown in the United Kingdom on this basis indicates that the breeder's task has been easiest for cereals, with potatoes and sugar beet considerably more difficult and herbage crops the most intransigent

TABLE 1. NATIONAL AVERAGE YIELDS (DRY MASS) AND TYPICAL HARVEST INDICES OF CROPS GROWN IN THE UNITED KINGDOM

	average yield, 1979 t/ha	harvest index†
wheat	4.4	44–48
barley	3.5	48–52
potatoes	6.6	75–85
sugar beet		
tap root	7.6	73
sugar	5.7	55
perennial ryegrass	no data	85–90

† Ratio of harvested dry matter to total dry matter, excluding fibrous roots.

TABLE 2. VARIETAL IMPROVEMENT IN YIELD OF CROPS GROWN IN ENGLAND AND WALES

	baseline variety	yield of best current varieties (percentage of baseline)	source of data
wheat	Bersee 1947	156	Silvey (1978)
barley	Plumage Archer 1947	132	
potatoes			
early	Home Guard 1943	120	Plant Breeding Institute trials (1980)
main crop	King Edward 1902	114	
perennial ryegrass	S24 1937	106	National Institute of Agricultural Botany
	S23 1933	108	(1980), *Recommended varieties of grasses*
sugar beet	no reliable data		

(tables 1 and 2). Thus the potato varieties Home Guard and King Edward are still recommended by the N.I.A.B., though they are not now grown on such large areas as varieties of higher yielding ability, notably the maincrop varieties Maris Piper and Pentland Crown and the early potato Maris Bard. In the Netherlands, Bintje, bred in 1910, is still the most widely grown variety despite large breeding programmes in which about one million seedlings are assessed each year.

There are no reliable estimates of genetic gain in yielding ability for sugar beet because the old varieties are no longer available. It is relevant that the rate of increase in national average yield for this crop kept pace with that of the cereals until the late 1960s, and it is now more difficult to obtain further advances in yield by selection within the genetic variability at present available to the breeder. Introduction of the monogerm character was an essential step in mechanization, and selection for greater vernalization requirement, to improve resistance to bolting, was a major contributor to yield increases from 1950 to 1970, enabling growers to bring sowing dates forward and thereby to extend the growing season by 4–6 weeks.

PROTECTION FROM HAZARDS

Alleviation of adverse climatic factors, and improvements in resistance to pests and diseases, are often of greater importance in the quest for higher and more stable production than are the genetic gains in yield potential that have been demonstrated for many crops.

Adaptation to soils and climate

For the small-grain cereal species, improved standing ability is probably the most significant achievement in overcoming limitations of the physical environment, even though it is of major value only where the soils and climate allow high yields to be obtained. Throughout the first half of this century, resistance to lodging in wheat was sought most strongly in west Europe, where considerable progress was made by exploiting polygenic variation for shorter, stronger straw. This improvement was, however, obtained at the expense of discarding very high proportions of breeding lines that did not meet the required standards, and may have retarded the rate of progress in yield potential. In this breeding situation one of the main attractions of major semi-dwarfing genes is in allowing breeders to give more attention to other characters.

Major semi-dwarfing genes were used by Japanese wheat breeders for many years before Salmon took seed of Norin 10 and related Japanese varieties to the U.S.A. in 1946 (Dalrymple 1980). This material was used by Vogel in breeding of Gaines and other semi-dwarf winter wheat varieties for the high-yielding areas of the Pacific Northwest. It enabled Borlaug to make rapid progress in breeding spring wheats for areas of Mexico and other low-latitude countries where irrigation was available or rainfall was adequate. In these countries comparatively little attention had previously been given to reducing height, because the soils were so infertile that lodging was rarely an important limiting factor. Against this background, the new semi-dwarf varieties provided the stimulus for rapid advances in agronomy of the crop, especially in the greater use of nitrogenous fertilizers and in the development of irrigation systems (Arnold 1980).

The distribution of the two major semi-dwarfing genes from Norin 10, *Rht*1 and *Rht*2, has since been determined for a wide range of varieties based on this material. These genes have also been shown to have advantageous pleiotropic effects on harvest index, grain number per spikelet and grain yield (Gale 1978). In west Europe, *Rht*1 and *Rht*2 have been used most intensively by the Plant Breeding Institute and by the Miln Marsters Group. Similar work in central and east Europe is based on other factors derived from Japanese varieties by the Italian breeder Strampelli. These factors have not been fully characterized, but it has been shown that they differ in their mode of action from *Rht*1 and *Rht*2 in being sensitive to gibberellic acid. They were also involved in the breeding of Talent, at present the most widely grown variety of wheat in France, and in some new lines recently entered into National List Trials by the Plant Breeding Institute.

In India the introduction of semi-dwarf varieties of wheat in the mid-1960s, combined with increases in agronomic inputs and a greater area sown to wheat, has led to increases in production comparable to those in west Europe, from 10–12 Mt in the years 1961–5 to 35 Mt in 1979.

THE ACHIEVEMENTS OF CONVENTIONAL BREEDING

This achievement has been criticized by some because the increase in area of wheat has been obtained partly at the expense of less productive crops, including legumes, but it is now more generally held that the considerable increase in overall production is nutritionally of much greater significance (Arnold 1980). Moreover, improvements in soil fertility, increases in the irrigated area and the introduction of earlier-maturing varieties have made it possible to extend the practice of multiple cropping. There are, however, wide disparities in yields of wheat between the States of India. Average yields in the Punjab, where irrigation is widely available, are now about three times those in States where the crop is rainfed (Swaminathan 1977). The yields from demonstration plots on trial grounds with irrigation and optimum fertilizer use confirm that water supply is the overriding factor. It is also clear that the prospects for advances in yield, at least in the short term, are greatest in those areas that have already benefited.

Plant breeding has had little success in increasing the yields of wheat in other ecologically difficult areas, especially where the effects of drought are accentuated by the temperature extremes of continental climates. For example, on the high plateau of Turkey and Iran the effective growing season is very short owing to low winter temperatures, and in North Africa the sudden onset of very high temperatures when the grain is filling frequently causes premature death of the plant. In such areas the crop is unreliable and average yields often show virtually no advance. The increases in production that have been obtained owe more to improvements in agronomy, especially in the management of fallows to conserve moisture, than to plant breeding.

There is a comparable situation with rice, where semi-dwarf varieties have led to greatly increased production in regions with well controlled irrigation systems, most notably in Thailand, North India and Sri Lanka. In these areas, where advantage can be taken of their ability to respond effectively to fertilizers, the potential for further increase in production has been estimated at 46% (Anon. 1978). About 70% of the world's rice is, however, rainfed and even in some monsoon regions the crop may suffer from drought in the later stages of development. For these areas, and on the plains of east India and Bangladesh, which do not have adequate flood water control, there has been little increase in average yields; the corresponding estimate for further improvement is only 3-8%.

Although there are many cases of genetic improvement in tolerance of low or high temperature, the most productive achievements in minimizing losses due to exigencies of the climate are probably those that have involved changes in the timing of developmental and growth stages to match periods of favourable weather. Extension of the environmental range of a crop species to new regions has often depended in the first instance on adaptation in this respect, followed by improvements in the distribution of assimilates as a second phase of the work. Water supply remains the principal limitation to yield in many situations, and even at the high levels of yield for arable crops common in west Europe, it is often the most important factor. In dryland areas the yields of some perennial crops have been increased by an ability to survive and resume growth after a period of drought. With annual crops, especially cereals, the predominant trend has been in avoidance of drought by earlier ear emergence and maturity. Further changes in this direction are, however, likely to be counter-productive owing to reductions in biomass that can be offset only partly by increases in harvest index (Fischer 1979).

Within species, there is little evidence of genetic gain in the efficiency of water use, as measured by the ratio of biomass to water transpired, and the resolution of this problem is of the highest priority. Reducing transpiration by increasing stomatal resistance to diffusion generally has a similar effect on uptake of CO_2, so mechanisms are needed that increase the diffusion gradient for CO_2 within the leaf. By analogy with drought-resistant species, these may have an anatomical basis in smaller mesophyll cells, or a biochemical one in CO_2 fixation, and identify closely with those proposed for higher photosynthetic rate. Thus, selection for greater biomass under conditions of water stress may also provide useful parental material for higher photosynthetic rate when water supply is not limiting (Austin 1980).

Disease resistance

A large proportion of the work in most breeding programmes is concerned with resistance to pests and disease, and may be overruling when alternative methods of chemical or husbandry control are too costly, or not sufficiently effective. A broad division may be made between problems in circumventing the ability of pathogens to evolve new virulences and work aimed at increasing the level of resistance. The former if typically associated with the obligate foliar pathogens of cereals, which tend to show host-specificity especially to species and to varieties, and the latter with facultative pathogens, which show specificity towards different species much more than towards different varieties (Scott *et al.* 1980).

The strategy of breeding for resistance to obligate pathogens and the progress obtained may be illustrated with yellow rust (*Puccinia striiformis*) of wheat. Very high levels of resistance are common, but there is no certain method of selecting breeding lines for durable resistance, defined as resistance that remains effective while a cultivar is widely grown (Johnson 1981). More than 10 new races of yellow rust with additional virulences on varieties or breeders' lines have been discovered in the United Kingdom over the last 15 years. Many of the varieties affected had major genes for race-specific resistance, usually seen as a hypersensitive reaction to infection in seedling tests. The genes concerned include *Yr1* to *Yr7* from *T. aestivum* varieties (Lupton & Macer 1962; Macer 1966), *Yr8* from *Aegilops comosa* as a chromosome 2M/2D translocation (Riley *et al.* 1968) and *Yr9* as a 1R/1B substitution from rye. Moreover, the deliberate assembly of combinations of such genes has not given a useful increase in durability. For this reason it is now common practice to avoid lines that have seedling-resistance to all known races.

There is indisputable evidence that durable resistance to yellow rust does exist, in that the resistance of some old varieties, notably Browick, Little Joss, Yeoman, Vilmorin 27, Atle, Hybride de Bersee, Cappelle-Desprez and Maris Widgeon, remained effective over long periods of cultivation. It also seems likely that Maris Huntsman has a residual level of durable resistance equivalent to Cappelle-Desprez. These varieties are all susceptible to one or more races in seedling tests but develop sufficient resistance in the adult plant to restrict the development of epidemics; this is sometimes referred to as 'slow-rusting'. However, selection for a combination of seedling susceptibility and adult plant resistance does not in itself guarantee durability, as is evident from the discovery of adult plant adapted variants of yellow rust on Joss Cambier in 1971, and on Maris Bilbo and Hobbit in 1973.

Although it is not possible to test for the durability of resistance to yellow rust, the probability of obtaining such resistance should be increased by the procedures described by Johnson (1978) and applied to the breeding of the winter variety Bounty (Bingham *et al.* 1980; Bingham 1981). Bounty was selected from a cross of an adult plant susceptible variety, Durin, with Ploughman; the latter was derived from a partial backcross of Hybrid 46 to Maris Widgeon as recurrent parent, and was therefore assumed to carry durable adult plant resistance. From the parents it would have been possible to combine race-specific genes giving complete resistance to all races known at the time, i.e. *Yr1*, *Yr3b* and *Yr4b* (table 3). This was avoided by selection for the same pattern of seedling resistance as Durin, so that adult plant resistance from Plougham could be detected.

Similar considerations apply to other obligate pathogens of the cereals, especially brown rust (*Puccinia recondita*), stem rust (*Puccinia graminis*) and powdery mildew (*Erysiphe graminis*). The work is hampered by a lack of information on the resistance mechanisms that determine

TABLE 3. PEDIGREE OF THE WINTER WHEAT VARIETY BOUNTY, SHOWING THE GENES FOR SEEDLING REACTION TO YELLOW RUST, AND THE REACTION OF BOUNTY AND ITS PARENTS TO TWO RACES OF YELLOW RUST IN THE SEEDLING AND ADULT PLANT STAGES

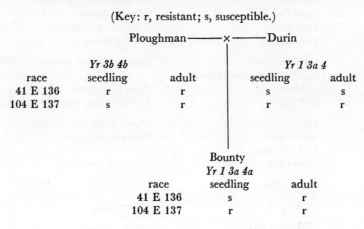

durability and their genetic control. Durability is not necessarily dependent on polygenic control (Eenink 1977), although it has been shown that some particular cases of durability are associated with a combination of more than one resistance mechanism (Russell 1978). No distinction has yet been made, however, between forms of adult resistance that have proved to be durable and those that give similar reactions in field tests, for example 'slow rusting', but are not durable.

Despite these limitations, many obligate pathogens have been kept under good control by breeding for resistance, one of the best known examples being resistant to stem rust of wheat in Australia, and also in North America where there have been no major epidemics since 1954. Such achievements have, however, been made at the expense of progress in other characters, owing to the high breeding input needed to maintain equilibrium with increasing virulence, and the obligation to discard many lines when new races appear. Many schemes have been proposed for a more orderly exploitation of resistance factors. Of these, the *Varietal diversification schemes* published by the National Institute of Agricultural Botany (Anon. 1981) are widely followed by growers, and have made a major contribution to the control of wheat yellow rust and barley mildew in the United Kingdom. Multilines and varietal mixtures are also complementary to pure line breeding, mixtures having the benefit of greater flexibility in composition, so that limitations in the range and combination of virulences possessed by pathogen populations can be continually exploited. This method is applicable to the control of many diseases where the durability of resistance is elusive, and its effectiveness has been clearly demonstrated in field trials with mixtures of spring barley varieties differing in genes for resistance to powdery mildew

(Wolfe *et al.* (1981). Mixtures could also be used to improve the stability of wheat yields in low-latitude countries, where it is most difficult to obtain durable resistance to brown rust.

In these ways some progress has been made in the strategy of using race-specific resistance and in methods of breeding for durability. There is no certainty that resistance introduced from alien sources will be more durable; this possibility should be investigated, but the main objective of new methods of genetic manipulation should be to improve the level of resistance to those pests and diseases where the sources of resistance available by conventional methods are in-

TABLE 4. EXAMPLES OF PESTS AND DISEASES OF ARABLE CROPS WITH SERIOUS LIMITATIONS IN THE LEVEL OF RESISTANCE ATTAINED BY CONVENTIONAL BREEDING METHODS

crop	pest/disease	organism
wheat	take-all	*Gaeumannomyces graminis*
	shoot fly	*Opomyza florum*
	bulb fly	*Leptohylemyia coarctata*
	† grain aphid	*Sitobion avenae*
barley	take-all	*Gaeumannomyces graminis*
	† barley yellow dwarf virus	
	‡ mildew	*Erysiphe graminis*
	aphids	*Sitobion avenae* and *Metopolophium dirhodum*
potato	gangrene	*Phoma exigua* var. *foveata*
	‡ late blight	*Phytophthora infestans*
	leaf roll virus	
	cream cyst nematode	*Globodera pallida*
sugar beet	† virus yellows	
oilseed rape	stem canker	*Phoma lingam*
field bean	chocolate spot	*Botrytis fabae*
	leaf and pod spot	*Ascochyta fabae*
	blackfly	*Aphis fabae*

† Moderate levels of resistance available but higher levels would be beneficial.
‡ Level of durable resistance not adequate.

adequate (table 4). Some of these objectives can be attained by established cytological methods (Law 1981); for example, resistance of wheat to eyespot (*Pseudocercosporella herpotrichoides*) has already been increased by transfer of additional factors from *Aegilops ventricosa* (Doussinault *et al.* 1974).

Technological properties and nutritional value

As the yields of many crops have increased, often to the point of surpluses in production, breeders have given increasing attention to the quality of the harvested product in terms of cooking quality, palatability and nutritional value, or suitability for manufacturing processes. In wheat the main objectives have been related to the technological requirements for the manufacture of bread, biscuits and other foods; little attention has been given to the possible effects of variety on the nutritional value of the grain, though there are many regulations governing the use of additives in manufacturing processes and the nutritional value of the foods produced.

The maritime climate of the United Kingdom is very well suited to the production of biscuit-making wheats, but difficult for bread-making quality, as has been well appreciated since the beginning of this century when all the varieties grown, such as Browick and Squarehead, were derivatives of land races of biscuit-making type. R. H. Biffen was the first to demonstrate that bread-making quality was heritable, by the breeding of Yeoman, marketed in 1916, from a cross of Browick with the Canadian spring wheat Red Fife. Bread-making quality was further improved by F. L. Engledow, using another Canadian variety, White Fife, in the parentage of Holdfast (1935), and these varieties figure in the lineage of Maris Widgeon (1964), Maris Freeman (1974), Bounty (1979) and Avalon (1980).

The United Kingdom has been self-sufficient in production of wheat for all uses other than bread-making since 1978. Increases in the area sown to varieties suitable for bread-making, in conjunction with improvements in baking technology, have enabled millers to reduce to 40–45 % the proportion of imported wheats of Manitoba type required in grists for white bread. However, imports of wheat in this class still amount to 1.8–2.0 Mt annually, so it is important to define the objectives for further varietal improvement. The best of present varieties meet all the requirements for the purely mechanical operation of milling, giving high flour-extraction rates. They also have the essential biochemical attributes of a low α-amylase activity in the grain, provided germination in the ear has not commenced, and good protein quality as shown by measurements of dough strength and elasticity. There are also excellent prospects for obtaining further improvements in protein quality by the deliberate recombination of glutenin subunits associated with good bread-making quality (Payne et al. 1981). The main breeding problems are concerned with liability to sprouting during wet harvest weather and with low grain protein content. It should be possible to obtain a sufficient improvement in harvest dormancy by the exploitation of recently discovered variation in *T. aestivum* but there are major physiological limitations to breeding for higher grain protein content.

There is a very strong tendency for varietal differences in grain protein content to be inversely related to yielding ability (Pushman & Bingham 1976). This is a general situation for many other crop plants, and arises largely because the opportunities for breeding varieties able to scavenge more nitrogen from the soil and increase harvest index for proteins are much less than those for transferring more carbohydrate products to the grain. Some genotypes indicate that it may be possible to offset in part the decline in protein content of higher-yielding varieties, but there is no prospect that a major improvement in this respect can be achieved by conventional breeding methods.

Nutritional value

Thus, for many crop plants, quality is defined by technological characters, and by consumer preferences in appearance, cooking properties and taste, which bear little relation to nutritional value. In fruit and vegetable crops especially, these objectives can be of greater importance than any other breeding character. The concept that nutritional value can be improved by breeding is of relatively recent origin, largely stemming from the discovery of the high lysine maize mutants *opaque 2* and *floury 2* (Mertz et al. 1964). No varieties of maize with this character are being grown on a substantial scale owing to associations with lower yielding ability, poorer cooking quality and greater susceptibility to storage pests and diseases, though some breeders are confident that these problems are surmountable.

Breeding for higher protein content involves a penalty in yielding ability partly because the energy requirement for the production of storage proteins, from a nitrate substrate, is practically double that needed for a similar mass of starch. There are only small differences between amino acids in this respect, so it should be possible to increase the proportion of limiting amino acids with no loss in yield. In practice, this objective has proved difficult to attain; many breeders have discontinued work with the high lysine barley mutant Risø 1508, owing to an association with shrunken grain, but other sources, especially from Ethiopian material, are considered more promising.

The elimination of toxic substances in rapeseed oil and meal is one of the most effective achievements in breeding for improved nutritional value. After the discovery of lines with low erucic acid content in Canada in 1961 (Downey 1974) and the implication in 1970 that this long-chain fatty acid was a possible cause of myocardial lesions, breeders were able to respond so rapidly that by 1980 an almost complete change to varieties with low erucic acid content had occurred. Feeding experiments have confirmed the improved nutritional value of such varieties, though it is now considered that the association of erucic acid content with myocardial lesions was originally exaggerated. Good progress is now being made in reduction of glucosinolates, which give breakdown products of bitter taste in rapeseed meal and reduce its consumption by animals.

Conclusion

Despite the wide differences between crop species in the environments where they are grown, in the pests and diseases that confront them, and in the plant organ that composes the harvested product, it is possible to identify some general limitations in objectives that can be attained by conventional breeding methods.

It is inevitable that genetic advances in yield potential will become increasingly difficult to obtain, and for many crops, especially those where harvest index is already high, the main limitations will be in biomass. There is probably little scope or advantage for increasing the leaf areas of high yielding crops grown in favourable environments. There may be some possibility for small advances in biomass by changes in leaf arrangement and in leaf longevity, but the greatest challenge is to increase photosynthetic rate per unit area of leaf. An increase in this respect would, however, be of limited value unless achieved by an anatomical or biochemical mechanism that also resulted in an equivalent increase in the efficiency of water use, thus giving improved resistance to drought.

There are two outstanding requirements in breeding for resistance to pests and diseases. One is for combined studies of the mechanisms and genetics of durable resistance, with the objective of devising selection methods applicable to large numbers of progenies. Some progress is, however, being made in strategies for handling available sources of resistance; the main limitation is often in the level of resistance that can be attained. New methods of genetic manipulation should therefore be directed primarily to increasing the level of resistance, by introduction of genetic factors from alien material not available by conventional methods.

Most consumer preferences, and the technological requirements for processing, can be met by breeding methods currently employed. There are also opportunities for improving nutritional value by changes in biochemical composition that are compatible with yield, but in many crops other than leguminous species there are serious restrictions in protein content.

REFERENCES (Bingham)

Anon. 1978 *Research highlights for 1977.* Manila: The International Rice Institute. (122 pages.)
Anon. 1981 *Recommended varieties of cereals.* Cambridge: National Institute of Agricultural Botany.
Arnold, M. H. 1980 Plant breeding. In *Perspectives in world agriculture*, pp. 67–89. Farnham Royal: Commonwealth Agricultural Bureau.
Austin, R. B. 1978 Actual and potential yields of wheat and barley in the United Kingdom. *ADAS Q. Rev.* 29, 76–87.
Austin, R. B. 1980 Physiological limitation to cereal yields and ways of reducing them by breeding. In *Opportunities for increasing crop yields* (ed. R. S. Hurd, P. V. Biscoe & C. Dennis, pp. 3–19. London: Pitman.
Austin, R. B., Bingham, J., Blackwell, R. D., Evans, L. T., Ford, M. A., Morgan, C. L. & Taylor, M. 1980 Genetic improvements in winter wheat yields since 1900 and associated physiological changes. *J. agric. Sci., Camb.* 94, 675–689.
Bingham, J. 1979 Wheat breeding objectives and prospects. *Agric. Prog.* 17, 1–17.
Bingham, J. 1981 Breeding wheat for disease resistance. In *Strategies for the control of cereal disease* (ed. J. F. Jenkyn & R. T. Plumb), pp. 3–14. Oxford: Blackwell Scientific Publications.
Bingham, J., Blackman, J. A. & Angus, W. J. 1980 *Wheat, a guide to varieties from the Plant Breeding Institute.* Newton, Cambridge: National Seed Development Organisation. (40 pages.)
Dalrymple, D. G. 1980 Development and spread of semi-dwarf varieties of wheat and rice in the United States. *United States Department of Agriculture, Agricultural Economic Report* no. 455. (150 pages.)
Doussinault, G., Koller, J., Touvin, H. & Dosba, F. 1974 Utilisation des géniteurs VPM 1 dans l'amélioration de l'état sanitaire du blé tendre. *Annls Amél. Pl.* 24, 215–41.
Downey, R. K. 1974 Breeding for modified fatty acid composition. *Curr. Adv. Pl. Sci.* 6, 151–167.
Eenink, A. H. 1977 Genetics of host–parasite relationships and the stability of resistance. In *Induced mutations against plant diseases*, pp. 47–57. Vienna: International Atomic Energy Authority.
Evans, L. T. & Dunstone, R. L. 1970 Some physiological aspects of evolution in wheat. *Aust. J. biol. Sci.* 23, 725–41.
Fischer, R. A. 1979 Growth and water limitation to dryland wheat yield in Australia: a physiological framework. *J. Aust. Inst. agric. Sci.*, pp. 83–94.
Gale, M. D. 1978 The effects of Norin 10 dwarfing genes on yield. In *Proc. 5th Int. Wheat Genetics Symp.*, New Delhi, India, pp. 978–987.
Johnson, R. 1978 Practical breeding for durable resistance to rust diseases in self-pollinating cereals. *Euphytica* 27, 529–40.
Johnson, R. 1981 Durable disease resistance. In *Strategies for the control of cereal disease* (ed. J. F. Jenkyn & R. T. Plumb), pp. 55–63. Oxford: Blackwell Scientific Publications.
Laabassi, M. A. L. 1979 Genetic male sterility in wheat: cytogenetic analysis and application to breeding procedures. Ph.D. thesis, University of Cambridge.
Law, C. N. 1981 Chromosome manipulation in wheat. In *Chromosomes today*, vol. 7 (ed. M. D. Bennett, M. Bobrow & G. M. Hewitt), pp. 194–205. London: George Allen & Unwin.
Lupton, F. G. H. & Macer, R. C. F. 1962 Inheritance of resistance to yellow rust (*Puccinia glumarum* Erikss. and Henn.) in seven varieties of wheat. *Trans. Br. mycol. Soc.* 45, 21–45.
Macer, R. C. F. 1966 The formal and monosomic genetic analysis of stripe rust (*Puccinia striiformis*) resistance. In *Proc. 2nd Int. Wheat Genetics Symp.* (*Hereditas, Suppl.* 2), pp. 127–42.
Mertz, E. T., Bates, L. S. & Nelson, O. E. 1964 Mutant gene that changes protein composition and increases lysine content of maize endosperm. *Science, N.Y.* 145, 279.
Payne, P. I., Corfield, Kathryn G., Holt, L. M. & Blackman, J. A. 1981 Correlations between the inheritance of certain high-molecular-weight subunits of glutenin and bread-making quality in progenies of six crosses of bread wheat. *J. Sci. Fd Agric.* 32, 51–60.
Pushman, F. M. & Bingham, J. 1976 The effects of a granular nitrogen fertiliser and a foliar spray of urea on the yield and bread-making quality of ten winter wheats. *J. agric. Sci., Camb.* 87, 281–92.
Riley, R., Chapman, V. & Johnson, R. 1968 The incorporation of alien disease resistance in wheat by genetic interference with the regulation of meiotic chromosome synapsis. *Genet. Res., Camb.* 12, 199–219.
Russell, G. E. 1978 *Plant breeding for pest and disease resistance.* London: Butterworths. (485 pages.)
Russell, W. A. 1974 Comparative performance for maize hybrids representing different eras of maize breeding. In *29th Annual Corn and Sorghum Research Conference*, pp. 81–101. Washington: American Seed Trade Association.
Scott, P. R., Johnson, R., Wolfe, M. S., Lowe, H. J. B. & Bennett, F. G. A. 1980 Host-specificity in cereal parasites in relation to their control. *Appl. Biol.* 5, 349–393.
Silvey, V. 1978 The contribution of new varieties to increasing cereal yield in England and Wales. *J. natn. Inst. agric. Bot.* 14, 367–384.

Swaminathan, M. S. 1977 Genetic and breeding research in wheat – the next phase. In *Proceedings of the 1st National Seminar on Genetics and Wheat Improvement*, Ludhiana, February 1977, pp. 3–20. Oxford and IBH Publishing Co.

Wolfe, M. S., Barrett, J. A. & Jenkins, J. E. E. 1981 The use of variety mixtures for disease control. In *Strategies for the control of cereal disease* (ed. J. F. Jenkyn & R. T. Plumb) pp. 73–80. Oxford: Blackwell Scientific Publications.

Discussion

P. S. WELLINGTON (*National Institute of Agricultural Botany, Cambridge, U.K.*). Skill in farm management could be a limiting factor that is now contributing to the increasing gap between the national average yield of cereals in the U.K. and the highest yields obtained by certain individual farmers. Reference was made in the paper to the estimates made by Mrs Silvey in 1978, that new varieties of wheat had contributed 50%, and new varieties of barley 30%, to the increases in national yields over the last three decades. Estimates were made at the same time of the contributions of other factors such as improved husbandry. It is significant that these made an additional contribution to national yields at least equal to that of new varieties in the period 1947–57, whereas improved techniques for cultivation and drilling and greater use of fertilizers, growth regulators, herbicides and fungicides did not appear to make any additional contribution between 1967 and 1977. It therefore seems that a high level of crop management is now required to obtain the higher yields from new varieties. This demands of the farmer detailed knowledge of local constraints, production planned to maximize profit and skilful use of the increasingly expensive inputs.

Reference was also made to the need for a more comprehensive system of variety trials to assess interactions with environmental conditions before new varieties are recommended to farmers. In practice the two main factors are sites and seasons, and the aim of the present N.I.A.B. trial system is to provide adequate samples of both for the main commercial production areas for the crop concerned. The recommendations are based on the statistical analysis of past trial results, but the farmer needs to predict the probable performance of the varieties, under his particular conditions, in a future season for which the weather conditions, in spite of long range forecasting, are not yet known. The present numbers of sites and seasons for variety trials are based on past experience, and calculation of the actual variations that have occurred. The objective is to use the resources available to obtain the best estimate of differences in the characteristics that make up commercial performance between all the varieties under consideration for inclusion on the Recommended Lists. The achievement of this objective requires complex methods of statistical analysis, but so far as seasonal variations are concerned, experience has shown that in practice the average over all sites for at least three, and if possible five, seasons provides a good prediction of behaviour under all except extreme conditions.

M. S. WOLFE (*Plant Breeding Institute, Cambridge, U.K.*). I should like to comment on the question of breeding for specific rather than general adaptation, and on the problem of defining, or predicting, the environmental variation that necessitates adaptability in the crop. In recent years, Dr J. A. Barrett and I have been investigating mixtures of cereal varieties, particularly barley, for disease control. Appropriately designed mixtures of three varieties have been highly effective in this respect, but we have become increasingly aware of the more general advantages of using mixtures over a wide range of environmental conditions. It is clear that they provide considerable buffering against both unpredictable and predictable environmental variation, so that stable high yields can be more easily attained, with fewer management problems for the farmer. Plants of different varieties grown side by side can compensate for each other under different environmental conditions; when they are grown in pure stands they cannot.

D. R. JOHNSTON (*Forestry Commission, Forest Research Station, Farnham, Surrey, U.K.*). Dr Wolfe has discussed buffering against biological factors. I should like to refer to buffering against economic trends. Plant breeders in forestry are beginning to question whether it is prudent to bequeath to posterity genetically improved tree crops that will take 50 years or so to mature and which have been selected and developed under high-input régimes. It is likely that the price of fertilizers and chemicals will increase in real terms, and it may be wise to devote some research effort to the breeding of genetic material that will be relatively successful under low-input régimes. I should be interested to know if this aspect of plant breeding has been considered by the agriculturists.

J. BINGHAM. In extensive areas of the world where the yields of cereals and other crops are severely restricted by limitations of soil or climate (especially water supply), fertilizers and other chemical inputs can be used economically only at relatively low levels. Much of the work of the international centres is now aimed at improving yields for such situations, including a major part of the programme at CIMMYT, International Centre for Maize and Wheat Improvement.

In the favourable climate of the U.K. it is more appropriate to select for efficiency of fertilizer use, as measured by the return in yield per unit of fertilizer applied. The increases in potential yield of varieties so far obtained are in the main related to improvements in this respect and are effective over a wide range in rate of N fertilizer use. There is a danger that such varietal differences may no longer be detected as the use of N fertilizer increases to very high rates. It is therefore desirable to reintroduce the practice of testing for yielding ability at more than one rate of N fertilizer application in national trials.

Induced gene and chromosome mutants

By R. A. Nilan

*Department of Agronomy and Soils and Program in Genetics,
Washington State University, Pullman, Washington 99164, U.S.A.*

Plant scientists, including breeders, can use an arsenal of physical and chemical mutagens and appropriate selection techniques to 'manufacture' in their experimental plots gene and chromosome mutants to compensate for the erosion of natural sources of genetic variability. They also have the capability of generating in this type of genetic manipulation the entire array of genetic variation inherent in all loci controlling each plant trait, and thus in a relatively short time producing most, if not all, of the genetic variants that have ever occurred in the evolution of a given agricultural plant.

This capability is required not only for the breeder concerned with developing new cultivars to meet the numerous and varied demands of the modern farmer, processor and consumer, but also for the geneticist, physiologist, anatomist and biochemist concerned with unravelling important plant processes and their genetic control. In short, these scientists need inexhaustible supplies of genetic variability, often never before selected in Nature or by earlier plant breeders.

Numerous experiments demonstrate that induced mutants have considerably extended the genetic variability of a phenotype. An outstanding example is *eceriferum* ('waxless' plant surfaces) in barley. Spontaneous mutations produced several well known variants controlled by about six loci. Genetic analyses of over 1300 induced and the few spontaneous mutants have determined that this trait is controlled by at least 77 loci (Lundqvist 1976, and personal communication). There are numerous alleles at some of these loci. Other examples are described in this paper.

The quantity and quality of artificially induced genetic variability in plants is in no small part due to the contributions of improved mutagens, mutagen treatments and selection techniques. A new potent and unique mutagen, sodium azide, is particularly successful in inducing putative point mutations. Recent experiments with barley and *Salmonella* have revealed that it is not azide *per se* but an activated metabolite that is the mutagenic agent. The metabolite has been isolated and crystallized and can now be synthesized *in vitro*. These findings usher in a new category of mutagens and suggest new avenues for understanding the interaction of mutagens with chromosomes and genes and for greater control of the induction of genetic variability in plants.

The considerable success of varietal development through induced mutants is well documented: 465 culvitars of sexually and vegetatively reproducing crops have been released that owe some of their production advantage to an induced gene or chromosome mutant. These cultivars have led to considerable economic impact in a number of countries.

In breeding research, induced mutants are indispensable for probing and elucidating the pathway and genetic control of important plant processes such as wax synthesis and deposition (von Wettstein-Knowles 1979), nitrogen assimilation (Kleinhofs *et al.* 1980), photorespiration and different facets of photosynthesis (Somerville & Ogren 1980; Miles *et al.* 1979; Simpson & von Wettstein 1980).

In the manipulation of plant genes (genetic engineering) in breeding research, it becomes increasingly necessary to pinpoint these genes on chromosomes. For this endeavour, an abundant array of induced chromosome mutants such as trisomics, telotrisomics, acrocentrics, inversions, translocations and deletions is required. This important activity can now be complemented by ever-improving chromosome banding techniques.

Introduction

Genetic variability, as derived to a greater extent from mutations comprising extragenic and possibly intragenic events (Nilan & Vig 1976; Brock 1980), and to a lesser extent from chromosome mutations (changes in chromosome number and structure), is the raw material of plant improvement. Until recently, genetic variability was secured by the plant breeder entirely from 'Nature's improvement programme' (rarely arising spontaneous mutations, subsequent recombination and natural selection) through collections of crop germ plasm, and wild species and even genera. In Nature, the genetic variants are end-products of thousands of years of evolution and were selected primarily for survival and reproductive capability.

Presumably during evolution, myriads of variants resulted from the variability potential of every locus and every chromosome break and many of the variants now needed by the plant breeder probably occurred. However, most were discarded since they were of little value for the plant's survival and fecundity. Evolutionary processes were not concerned with preserving the numerous traits that are now required from plants by the modern farmer for higher yield and adaptation to advanced farming practices and to new environmental niches; by the modern processor of food and fibre; and by the animal and human consumers who require an increased variety of foods with improved nutrition, palatability and attractiveness. These requirements are increasing while at the same time natural genetic variability in some crops is eroding.

The plant breeder needs now, and will do more so in the future, a broad array of genetic variation, possibly for every locus, whether it be a structural or a regulatory gene, for every plant process and phenotype. This is required most assuredly for the relatively few plants upon which we now base our production of food and fibre and for those plants that have the potential of broadening the food and fibre base and for fulfilling special needs. This genetic variation, plus all the conceivable changes in plant karyotypes that can be achieved through chromosome mutants, will eventually be required to reconstruct, so to speak, plants for man's needs. Obviously, the breeder cannot wait for new and usable spontaneous mutations.

Fortunately, plant geneticists and breeders, using an arsenal of physical and chemical mutagens and appropriate selection techniques, can now 'manufacture' gene and chromosome mutants to compensate somewhat for the depletion of natural sources of genetic variability. They also have the capability in this type of genetic manipulation of generating experimentally the entire array of genetic variation inherent in all loci controlling each plant trait, and thus in a relatively short time producing most, if not all, of the genetic variants that have ever occurred or may yet occur in the evolution of agricultural plants. The bases for the above concepts have been previously recounted (Brock 1980; Nilan et al. 1977).

This paper considers the current and future impact of induced gene and chromosome mutants on plant improvement. It also examines the impact of induced mutants on breeding research, especially towards providing new knowledge about the genetics, physiology, anatomy and biochemistry of cellular processes that produce all the traits so necessary for successful cultivars now and in the future.

Induced and spontaneous mutations

A breeder contemplating induced mutations might first ask, 'Do induced and spontaneous mutations and mutants differ?' and 'What is the value of induced mutants when they are so "raw" and have not been moulded by evolution and the various recombination and selection processes that have developed spontaneous mutants into useful adapted complexes?' It has been amply demonstrated in a wide variety of organisms, including plants, that there are actually no major detectable differences between induced and spontaneous gene and chromosome mutations (Brock 1980; Nilan *et al.* 1977; Nilan & Vig 1976; Konzak *et al.* 1977). Of interest to the breeder is that the older spontaneous mutants have been moulded by recombination and natural selective forces into useful co-adaptive complexes. Newly arisen spontaneous mutants have not had time to be moulded into such complexes. However, the artificial acceleration of recombination plus refined and discrete selection techniques in the hands of the plant breeder can soon lead to useful trait complexes whether the origin of the trait is by induced or spontaneous mutation. Indeed, Brock (1980), on several pieces of evidence, questions the value of co-adapted complexes in plant improvement.

Induced 'new' variability

Another question often asked by plant scientists contemplating the use of induced mutations is, 'Can induced mutations produce "new" forms of traits that have not been observed among the spontaneous genetic variability?' The answer is that they can. Artificial mutagens can produce mutants that have not arisen in recent evolutionary history and thus have never been encountered by the breeder. However, they are probably not 'new' to a given plant because such variants may have occurred during its evolution.

In plants such as barley and peas, and to a lesser extent maize and wheat, that have been used extensively in basic and applied mutagenesis research, the extension of genetic variability by induced mutations and mutants is well documented. In these examples, induced mutants have uncovered hitherto unknown or 'new' loci controlling a phenotype and have revealed much about the potential variability (alleles) of many loci.

One of the most striking examples of the induction of 'new' variability and loci for a phenotype is represented by variants for the *eceriferum* ('wax-less' plant surface) phenotype of barley. Natural variability for this trait was confined to a few spontaneous mutants at six controlling loci. By using a wide variety of mutagens, 1302 mutants for this trait have been induced (Lundqvist 1976; Lundqvist *et al.* 1968; Lundqvist, personal communication). These numbers of independent gene changes differentially affect the wax composition of the leaf blade and sheath, spike and stem. They also lead to remarkable and distinct differences in fine structure and chemical composition of the surface wax molecules (von Wettstein-Knowles 1976, 1979). Appropriate genetic tests of the induced and spontaneous mutants have revealed at least 77 loci mapped to each arm of the seven barley chromosomes and numerous alleles, over 100, occurring at each of several loci. Similarly, in barley, induced chlorophyll-deficient mutants have revealed 600–700 loci controlling chlorophyll development (von Wettstein *et al.* 1974; Nilan & Velemínský 1981; Simpson & von Wettstein 1980) and 26 loci, some with numerous alleles, for the *erectoides* trait (Persson & Hagberg 1969). Moreover, genetic variability has been greatly broadened through induced mutation techniques for such phenotypes in barley as anthocyanin development (Jende-Strid 1978), nitrate reductase (Kleinhofs *et al.* 1980), mildew resistance (Jørgensen 1976), spike development (Gustafsson & Lundqvist 1980), and for lysine content (Doll *et al.* 1974). That the examples above are in barley testifies to the fact that among all of the crop plants, and indeed higher plants, there has been no other plant that has received so much investigation in the area of mutagenesis. In short, it has been the plant model of choice for experimental mutagenesis and mutation breeding. Examples of how induced mutants have revealed 'new' loci in other plants have been given previously (Nilan *et al.* 1977; Konzak *et al.* 1977; Brock 1980).

Improved techniques

The increasing success of induced gene and chromosome mutants in breeding and breeding research can be attributed to improvements in mutation induction and selection techniques. These, along with relevant literature and descriptions of the most useful physical and chemical mutagens, recipes for their use on appropriate plant parts, e.g. seeds, buds, pollen, tissue and cells, and techniques for inducing and selecting mutants in sexually and vegetatively reproducing crops, are presented in the International Atomic Energy Agency's *Manual on mutation breeding* (1977).

There are numerous data and well documented technology that can lead to greater mutagen effectiveness (frequencies of mutations per dose of mutagen), efficiency (frequencies of desired events such as gene mutations in relation to such undesirable or unwanted events as sterility and, in some cases, chromosome aberrations), and specificity (group mutability (spectrum alteration), interlocus and non-random chromosome breakage) (Konzak *et al.* 1965; Nilan 1972; Brock 1980). With judicial selection of mutagens and manipulation of mutagen treatments, the breeder can influence the kind of genetic events that he may wish to induce as sources of genetic variability for his improvement programme. For instance, all of the physical mutagens such as X-rays, γ-rays and neutrons, as well as certain chemicals such as myleran, can induce high ratios of chromosome aberrations to mutations. There are other mutagens, e.g. ethyleneimine, that can induce about equal frequencies or proportions of both. Finally, there are mutagens such as ethyl methanesulphonate, diethylsulphate, sodium azide and certain base substitution and nitroso compounds that appear to induce higher proportions of mutations to chromosome aberrations (Konzak *et al.* 1977). As more knowledge is obtained about the mechanism, action and specificity of mutagens, and the nature of the mutations that they induce, the breeder will acquire even more precision for advantageously inducing and manipulating mutants in plant improvement.

At Pullman, 10 years of extensive basic research has developed a relatively new mutagen, sodium azide (Sideris *et al.* 1973; Nilan *et al.* 1973), which is one of the most potent available for higher plants. The research with this mutagen has been recently summarized by Kleinhofs *et al.* (1978a).

Azide is unique in that it induces in plants very high frequencies of gene mutations but is ineffective in producing major chromosomal changes. Experiments with bacteria indicate that azide is a base substitution mutagen, and in eukaryotes it appears to induce changes on the order of point mutations. Whether these mutations are small deletions or true base changes has not yet been resolved. We are attempting to answer this question by mapping numerous alleles induced by azide at the waxy pollen locus of barley (Rosichan *et al.* 1981; Nilan *et al.* 1981). Here the nature of the mutant alleles can be genetically resolved to near the base-pair level, since rare interallelic recombination events can be detected on a per million pollen basis. Preliminary results suggest that distances between alleles of about 50 base pairs can be detected, indicating that at least some mutational events do not involve large DNA deletions.

Recently, we have determined that it is not the inorganic azide *per se* but an organic metabolite synthesized in azide-treated barley and bacteria cells that is the mutagenic agent (Owais *et al.* 1978, 1979). This metabolite has been isolated, purified, crystallized and partly characterized chemically (Owais *et al.* 1981 *b*) and its *in vitro* synthesis from cell-free extracts of *Salmonella typhimurium* has been accomplished (Owais *et al.* 1981 *a*). Furthermore, the pathway by which this metabolite is synthesized is being revealed (Owais *et al.* 1981 *c*; Cieśla *et al.* 1980).

Although activation of numerous chemicals to mutagenic metabolites is well known in mammalian mutagenesis, very little research in this area has been conducted in plants. Indeed, the azide metabolite is one of only three (atrazine (Plewa & Gentile 1976) and 1,2-dibromoethane (Scott *et al.* 1978) being the others) that have been detected, and the only one in plants that has been isolated and purified to crystal form and about which knowledge of its synthesis is becoming available. It is now suspected that additional chemicals may act the same way in plants.

This type of research is providing a greater insight about the interactions of mutagens with genes and chromosomes and the nature of induced genetic change. It is also providing the breeder with new knowledge and technology with which he can 'manufacture', with considerable deliberation, greater genetic variability.

Induced mutants for cultivar improvement

There is now overwhelming evidence that induced mutants have contributed most significantly to breeding new cultivars of crops (Sigurbjörnsson 1976; Gustafsson 1975; Sigurbjörnsson & Micke 1974; Broertjes & Van Harten 1978; A. Micke, personal communication). By September 1980, at least 224 cultivars of self and cross-pollinating crop species had been released for commercial production around the world (table 1). These cultivars, possessing at

Table 1. Released induced mutant cultivars
(September 1980.)

type of crop	direct	cross
cereals	74 (total)	57 (total)
bread wheat	12	5
durum wheat	5	7
rice	28	9
barley	25	33
oats	4	3
legumes	18	10
fruit trees	8	1
other crops	46	10
total crops	146	78
ornamentals	237	4
total	**383**	**82**

After Sigurbjörnsson & Micke (1974), Sigurbjörnsson (1976), and A. Micke (personal communication).

least one improved trait due to an induced mutation, include 131 cereals, 28 legumes, and 9 fruit trees. In addition, 241 new strains of vegetatively reproducing species, mostly ornamental have been released. Cultivars that owe their advantage to induced mutants have been developed in 37 countries and grown successfully on millions of hectares, and thus have had considerable economic impact in numerous countries. In some countries induced mutant cultivars have enjoyed most of the acreage devoted to a given crop species.

The techniques for utilizing induced mutant genes for both qualitatively and quantitatively inherited traits and chromosome mutants in breeding have been adequately described in numerous reviews (for instance Brock 1980; Gaul 1964; Nilan et al. 1965) and publications from the International Atomic Energy Agency, especially the *Manual on mutation breeding* (1977).

In sexually propagating species, induced mutants can be used in two principal ways: directly, or in crosses or hybridization. In the former, a mutant that exhibits at least one improved trait with no new undesirable traits as a result of the induced genetic changes is multiplied directly. Once the mutant has been sufficiently tested with positive results, then it can be released to growers as a new cultivar. Among the 224 crop cultivars developed through induced mutants, 78 have been developed by direct multiplication of the mutant line (table 1). One advantage of the method is the short time required for developing a new cultivar. An example is the breeding of the six-row winter barley 'Luther' at Washington State University. Only 6 years elapsed from the time of mutagen treatment of seeds of the parent cultivar 'Alpine' to release of the mutant cultivar to growers in the Pacific Northwest of the U.S.A.

The plant breeder also can effectively use induced mutants in crosses – a necessity when the desired improved trait is closely linked or associated with undesirable spontaneous or induced traits. Furthermore, even directly useful mutant cultivars have proved to be outstanding parents for cross-breeding. The latter is well illustrated by barley. Of the 58 barley cultivars released with an induced mutant in their backgrounds, 33 have resulted from crossing mutants with other varieties or lines (table 1). Six successful cultivars were developed from the Swedish mutant cultivar Mari (Gustafsson 1975). In our barley breeding programme, the mutant cultivar Luther has been the parent of one released cultivar in Washington and of several advanced selections pending release in the states of Oregon, Washington, and Idaho.

In vegetatively propagated species (ornamentals, including cut flowers, bulbs, trees and shrubs; fruits; potato; sweet potato; sugar cane; cassava), much plant improvement has been based on the selection of 'sports' or spontaneous mutants. Thus, induced mutants are an obvious supplementary source of genetic variability. According to Broertjes & Van Harten (1978), the main advantage of mutation induction in vegetatively propagated plants is the ability to change one or a few characteristics of an outstanding genotype or cultivar without altering the remaining phenotype. In such plants, selection and propagation of useful mutants is relatively easy and development of mutant cultivars quite rapid.

The decision to use induced mutants in breeding will depend on the available supply of natural variability (and for some crops this is rapidly being depleted), the potential for success, and the effort and cost, especially where the utility of induced mutants are compared with securing the needed variants from related species or genera. These aspects of mutation breeding have been thoroughly analysed and discussed by Brock (1971, 1980).

Induced mutants for breeding research

The molecular basis of the genetic, biochemical, physiological and anatomical processes leading to those traits that comprise a successful crop cultivar are little understood. Yet the plant breeder today, and especially in the future, must learn to control and manipulate these processes if new cultivars to meet the requirements of the farmer, processer and consumer are to be met. Some of the progress and problems in understanding these processes, which are in the realm of plant molecular biology, have been recently reviewed (Walbot 1980) and are described elsewhere in this symposium.

In microorganisms, and even certain well studied animal species, one of the requirements for advancing knowledge about the molecular bases of cell processes is an array of mutant lines, often induced, that modify or block steps in the process under study. Until recently, this approach has been neglected in plants and probably accounts for the lack of knowledge and slow development of technology in plant molecular biology.

Examples of the role of induced mutants in probing developmental and cellular processes and their genetic control use a broad spectrum of induced mutants with specific defects. Some facets of this approach have been reviewed by Rice & Carlson (1975) and Scholz & Böhme (1980). The former also present some valuable ideas about the use of induced mutants in analysing seed development and relevant biochemical and physiological processes.

The use of numerous *eceriferum* ('waxless' plant surface) mutants (von Wettstein-Knowles 1979) in barley is elucidating the pathway of wax synthesis and deposition; mutants lacking in serine–glyoxylate aminotransferase activity in *Arabidopsis* are permitting an understanding of photorespiration and its genetic control and regulation (Somerville & Ogren 1980); detailed genetic, biochemical and ultrastructural analyses of innumerable chlorophyll-deficient mutants in barley (von Wettstein *et al.* 1974; Simpson & von Wettstein 1980), and of high chlorophyll fluorescence mutants in maize (Miles *et al.* 1979), are providing an understanding of the regulation, genetic control and metabolic pathways involved in various facets of photosynthesis; ten induced nitrate-reductase deficient mutants in barley are permitting biochemical, genetic and physiological investigations toward an understanding and control of the nitrate assimilation pathway (Kleinhofs *et al.* 1978b; Warner & Kleinhofs 1974; Kleinhofs *et al.* 1980); waxy pollen mutants in barley are being used to probe with classical and molecular genetic techniques the nature of induced mutations, the composition of a eukaryotic locus and the synthesis and deposition of starch, and to develop a mutagen monitoring system (Rosichan *et al.* 1981; Nilan *et al.* 1981; Hodgdon *et al.* 1981); and numerous induced anthocyanin-free mutants in barley are helping us to understand the pathway and genetic control of anthocyanin synthesis and to develop strains free of proanthocyanidins, which are responsible for permanent chill haze and instability in beer (von Wettstein *et al.* 1980; von Wettstein 1979; Jende-Strid 1978; R. A. Nilan & A. L. Hodgdon, unpublished).

Success in using induced mutants in cell cultures, protoplasts and pollen for analysing basic processes has been very limited. Recent developments, and the problems inherent in mutant induction and selection, and especially plant regeneration, have been reviewed (Rice & Carlson 1975; Brock 1980) and are described in more detail by Davies (this symposium).

Another important area of breeding research involves chromosome mutants and the location of genes on chromosomes. The efficient assembly of necessary genotypes for analyses of biochemical and physiological processes and progress in manipulating genes in breeding and breeding research (genetic engineering) requires that each gene or set of genes contributing to a trait be pinpointed on the chromosome. Success in this endeavour will require a vast array of induced chromosome mutants and improved chromosome banding techniques. In short, this area of cytogenetics should become as important for plant improvement as it has been for advancing knowledge and technology in human genetics and medicine.

Locating genes on a specific chromosome is facilitated by trisomics (Khush 1973; Lewis *et al.* 1980; Tsuchiya 1969), monosomics (Kimber & Sears 1980; Law *et al.*, this symposium) and translocation break points. The latter may be recognized and used in mapping through partial sterility (Ramage *et al.* 1961) or cytologically (Tuleen 1971). In barley, over 300 translocations, mostly induced, are available for cytologically locating genes (Nilan 1974). Induced translocation break-points can often provide cytological markers in chromosome regions lacking suitable genes.

To locate genes on a specific arm, telocentrics (Sears & Sears 1978; Kimber & Sears 1980; Singh & Tsuchiya 1977) and translocation break points are indispensable. In maize, numerous induced B–A translocations have been useful (Beckett 1978). To pinpoint genes cytologically within chromosome arms, translocation break-points as well as deletions, as so elegantly demonstrated in tomato by Khush & Rick (1968), and acrocentrics (Tsuchiya & Hang 1980), are necessary. Chromosome banding, now being used for locating genes (Kimber & Sears 1980; Linde-Laursen 1979) will be a powerful complementary tool in this endeavour.

Conclusion

Improved mutagen treatments, along with increased precision in selection of resulting mutants, are rapidly increasing the use and success of induced mutations in plant improvement. Such mutations are substituting for and even extending the variability obtained from natural germ plasm sources. As natural variability becomes further depleted and a much greater supply of variants is needed to create new cultivars of the future, artificially induced variability will assume greater importance. Indeed, the relative ease of producing and the suitability of induced variability for some crops may reduce or even negate the need for collection and preservation of natural germ plasm.

Induced gene and chromosome mutants are already proving indispensable for elucidating new basic knowledge about physiological, biochemical and genetic processes composing phenotypes and their control and for pinpointing genes on chromosomes.

Results attributed to me and my colleagues were obtained from investigations supported by the College of Agriculture Research Center, Washington State University, Pullman, Washington, Project nos 1006 and 1068, Department of Energy Contract no. DE-AM06-76RL02221, National Institute of Environmental Health Sciences Grant ES02224, and the Carlsberg Research Foundation, Copenhagen, Denmark, DOE/RL/02221-53.

References (Nilan)

Beckett, J. B. 1978 B–A translocations in maize. I. Use in locating genes by chromosome arms. *J. Hered.* **69**, 27–36.
Brock, R. D. 1971 The role of induced mutations in plant improvement. *Radiat. Bot.* **11**, 181–196.
Brock, R. D. 1980 Mutagenesis and crop improvement. In *The biology of crop productivity* (ed. P. S. Carlson), pp. 383–409. New York: Academic Press.
Broertjes, C. & Van Harten, A. M. 1978 *Application of mutation breeding methods in the improvement of vegetatively propagated crops.* Amsterdam: Elsevier Scientific Publishing Company.
Cieśla, Z., Filutowicz, M. & Klopotowski, T. 1980 Involvement of the L-cysteine biosynthetic pathway in azide-induced mutagenesis in *Salmonella typhimurium. Mutat. Res.* **70**, 261–268.
Doll, H., Køie, B. & Eggum, B. O. 1974 Induced high lysine mutants in barley. *Radiat. Bot.* **14**, 73–80.
Gaul, H. 1964 Mutations in plant breeding. *Radiat. Bot.* **4**, 155–232.
Gustafsson, Å. 1975 Mutations in plant breeding – a glance back and a look forward. In *Radiation research* (*Proc. Fifth Int. Cong. Radiat. Res.*) (ed. O. F. Nygaard, H. I. Adler & W. K. Sinclair), pp. 81–95. New York: Academic Press.
Gustafsson, Å. & Lundqvist, U. 1980 Hexastichon and intermedium mutants in barley. *Hereditas* **92**, 229–236.
Hodgdon, A. L., Marcus, A. H., Arenaz, P., Rosichan, J. L., Bogyo, T. P. & Nilan, R. A. 1981 The ontogeny of the barley plant as related to mutation expression and detection of pollen mutations. *Environ. Hlth Perspect.* **37**. (In the press.)

Jende-Strid, B. 1978 Mutations affecting flavonoid synthesis in barley. *Carlsberg Res. Commun.* **43**, 265–273.

Jørgensen, J. H. 1976 Identification of powdery mildew resistant barley mutants and allelic relationship. In *Barley genetics III* (ed. H. Gaul), pp. 446–455. Munich: Karl Thiemig.

Khush, G. S. 1973 *Cytogenetics of aneuploids.* New York: Academic Press.

Khush, G. S. & Rick, C. M. 1968 Cytogenetic analysis of the tomato genome by means of induced deficiencies. *Chromosoma* **23**, 452–484.

Kimber, G. & Sears, E. R. 1980 Uses of wheat aneuploids. In *Polyploidy: biological relevance* (ed. W. H. Lewis), pp. 427–443. New York: Plenum Publ. Corp.

Kleinhofs, A., Kuo, T. & Warner, R. L. 1980 Characterization of nitrate reductase-deficient barley mutants. *Molec. gen. Genet.* **177**, 421–425.

Kleinhofs, A., Owais, W. M. & Nilan, R. A. 1978a Azide. *Mutat. Res.* **55**, 165–195.

Kleinhofs, A., Warner, R. L., Muehlbauer, F. J. & Nilan, R. A. 1978b Induction and selection of specific gene mutations in *Hordeum* and *Pisum*. *Mutat. Res.* **51**, 29–35.

Konzak, C. F., Nilan, R. A. & Kleinhofs, A. 1977 Artificial mutagenesis as an aid in overcoming genetic vulnerability of crop plants. In *Genetic diversity in plants* (ed. A. Muhammed, R. Aksel & R. C. von Borstel), pp. 163–177. New York: Plenum Publ. Corp.

Konzak, C. F., Nilan, R. A., Wagner, J. & Foster, R. J. 1965 Efficient chemical mutagenesis. In *The use of induced mutations in plant breeding* (*Radiat. Bot., Suppl.* 5), pp. 49–70. Oxford: Pergamon.

Lewis, E. J., Humphreys, M. W. & Caton, M. P. 1980 Chromosome location of two isozyme loci in *Lolium perenne* using primary trisomics. *Theor. appl. Genet.* **57**, 237–239.

Linde-Laursen, I. 1979 Giemsa C-banding of barley chromosomes. III. Segregation and linkage of C-bands on chromosomes 3, 6 and 7. *Hereditas* **91**, 73–77.

Lundqvist, U. 1976 Locus distribution of induced *eceriferum* mutants in barley. In *Barley genetics III* (ed. H. Gaul), pp. 162–163. Munich: Karl Thiemig.

Lundqvist, U., Wettstein-Knowles, P. von & Wettstein, D. von 1968 Induction of *eceriferum* mutants in barley by ionizing radiations and chemical mutagens. II. *Hereditas* **59**, 473–504.

Manual on mutation breeding 1977 Tech. Reps Ser. no. 119, 2nd edn. Vienna: I.A.E.A.

Miles, C. D., Markwell, J. P. & Thornber, J. P. 1979 Effect of nuclear mutation in maize on photosynthetic activity and content of chlorophyll-protein complexes. *Pl. Physiol.* **64**, 690–694.

Nilan, R. A. 1972 Mutagenic specificity in flowering plants: Facts and prospects. In *Induced mutations and plant improvement*, pp. 141–151. Vienna: I.A.E.A.

Nilan, R. A. 1974 Barley (*Hordeum vulgare*). In *Handbook of genetics* (ed. R. C. King), vol. 2, pp. 93–100. New York: Plenum Press.

Nilan, R. A., Kleinhofs, A. & Konzak, C. F. 1977 The role of induced mutation in supplementing natural genetic variability. *Ann. N.Y. Acad. Sci.* **287**, 367–384.

Nilan, R. A., Konzak, C. F., Wagner, J. & Legault, R. R. 1965 Effectiveness and efficiency of radiations for inducing genetic and cytogenetic changes. In *The use of induced mutations in plant breeding* (*Radiat. Bot., Suppl.* 5), pp. 71–89. Oxford: Pergamon.

Nilan, R. A., Rosichan, J. L., Arenaz, P., Hodgdon, A. L. & Kleinhofs, A. 1981 Pollen genetic markers for detection of mutagens in the environment. *Environ. Hlth Perspect.* **37**. (In the press.)

Nilan, R. A., Sideris, E. G., Kleinhofs, A., Sander, C. & Konzak, C. F. 1973 Azide – a potent mutagen. *Mutat. Res.* **17**, 142–144.

Nilan, R. A. & Veleminský, J. 1981 Mutagenicity of selected chemicals in barley test systems. In *Proc. Comparative Chemical Mutagenesis Workshop* (ed. F. J. de Serres), ch. 11. (In the press.)

Nilan, R. A. & Vig, B. K. 1976 Plant test systems for detection of chemical mutagens. In *Chemical mutagens. Principles and methods for their detection* (ed. A. Hollaender), vol. 4, pp. 143–170. New York: Plenum Press.

Owais, W. M., Kleinhofs, A. & Nilan, R. A. 1979 In vivo conversion of sodium azide to a stable mutagenic metabolite in *Salmonella typhimurium*. *Mutat. Res.* **68**, 15–22.

Owais, W. M., Kleinhofs, A. & Nilan, R. A. 1981a Effects of L-cysteine and O-acetyl-L-serine in the synthesis and mutagenicity of azide metabolite. *Mutat. Res.* **80**, 99–104.

Owais, W. M., Kleinhofs, A. & Nilan, R. A. 1981b In vitro synthesis of the mutagenic metabolite by cell-free bacterial extracts. (Submitted.)

Owais, W. M., Kleinhofs, A., Ronald, R. C. & Nilan, R. A. 1981c Isolation of an azide mutagenic metabolite in *Salmonella typhimurium*. *Mutat. Res.* (In the press.)

Owais, W. M., Zarowitz, M. A., Gunovich, R. A., Hodgdon, A. L., Kleinhofs, A. & Nilan, R. A. 1978 A mutagenic in vivo metabolite of sodium azide. *Mutat. Res.* **50**, 67–75.

Persson, G. & Hagberg, A. 1969 Induced variation in a quantitative character in barley. Morphology and cytogenetics of *erectoides* mutants. *Hereditas* **61**, 115–178.

Plewa, M. J. & Gentile, J. M. 1976 Mutagenicity of atrazine: a maize–microbe bioassay. *Mutat. Res.* **38**, 287–292.

Ramage, R. T., Burnham, C. R. & Hagberg, A. 1961 A summary of translocation studies in barley. *Crop Sci.* **1**, 277–279.

Rice, T. B. & Carlson, P. S. 1975 Genetic analysis and plant improvement. *A. Rev. Pl. Physiol.* **26**, 279–308.

Rosichan, J., Arenaz, P., Blake, N., Hodgdon, A., Kleinhofs, A. & Nilan, R. A. 1981 An improved method for the detection of mutants at the waxy locus in Hordeum vulgare. *Environ. Mutagenesis.* **3**. (In the press.)

Scholz, G. & Böhme, H. 1980 Biochemical mutants in higher plants as tools for chemical and physiological investigations – a survey. *Kulturpflanze* **28**, 11–32.

Scott, B. R., Sparrow, A. H., Schwemmer, S. S. & Schairer, L. A. 1978 Plant metabolic activation of 1,2-dibromoethane (EDB) to a mutagen of greater potency. *Mutat. Res.* **49**, 203–212.

Sears, E. R. & Sears, L. M. S. 1978 The telocentric chromosomes of common wheat. In *Proc. 5th Int. Wheat Genet. Symp.*, New Delhi, pp. 389–407.

Sideris, E. G., Nilan, R. A. & Bogyo, T. P. 1973 Differential effect of sodium azide on the frequency of radiation-induced chromosome aberrations vs. the frequency of radiation-induced chlorophyll mutations in *Hordeum vulgare*. *Radiat. Bot.* **13**, 315–322.

Sigurbjörnsson, B. 1976 The improvement of barley through induced mutation. In *Barley genetics III* (ed. H. Gaul), pp. 84–95. Munich: Karl Thiemig.

Sigurbjörnsson, B. & Micke, A. 1974 Philosophy and accomplishments of mutation breeding. In *Polyploidy and induced mutations in plant breeding*, pp. 303–343. Vienna: I.A.E.A.

Simpson, D. J. & von Wettstein, D. 1980 Macromolecular physiology of plastids. XIV. *Viridis* mutants in barley: genetic fluoroscopic and ultrastructural characterisation. *Carlsberg Res. Commun.* **45**, 283–314.

Singh, R. J. & Tsuchiya, T. 1977 Morphology, fertility, and transmission in seven monotelotrisomics of barley. *Z. PflZücht.* **78**, 327–340.

Somerville, C. R. & Ogren, W. L. 1980 Photorespiration mutants of *Arabidopsis thaliana* deficient in serine-glyoxylate aminotransferase activity. *Proc. natn. Acad. Sci. U.S.A.* **77**, 2684–2687.

Tsuchiya, T. 1969 Status of studies of primary trisomics and other aneuploids in barley. *Genetica* **40**, 216–232.

Tsuchiya, T. & Hang, A. 1980 Acrocentric chromosome $4S^{4L}$ in barley. *Barley Genet. Newsl.* **10**, 72–73.

Tuleen, N. A. 1971 Translocation-gene linkages from F_2 seedlings in barley. In *Barley genetics II* (ed. R. A. Nilan), pp. 208–212. Pullman: Washington State University Press.

Walbot, V. 1980 Molecular biology of higher plants. In *The biology of crop productivity* (ed. P. S. Carlson), pp. 343–409. New York: Academic Press.

Warner, R. L. & Kleinhofs, A. 1974 Relationships between nitrate reductase, nitrite reductase, and ribulose diphosphate carboxylase activities in chlorophyll-deficient mutants of barley. *Crop Sci.* **14**, 654–658.

Wettstein, D. von 1979 Biochemical and molecular genetics in the improvement of malting barley and brewers yeast. In *Proc. 17th Congr. European Brewery Convention*, West Berlin, pp. 587–629.

Wettstein, D. von, Jende-Strid, B., Ahrenst-Larsen, B. & Erdal, K. 1980 Proanthocyanidin-free barley prevents the formation of beer haze. *MBAA tech. Q.* **17**, 16–23.

Wettstein, D. von, Kahn, A., Nielsen, O. F. & Gough, S. 1974 Genetic regulation of chlorophyll synthesis analyzed with mutants in barley. *Science, N.Y.* **184**, 800–802.

Wettstein-Knowles, P. von 1976 Tracking down β-diketone synthesis with the aid of the *eceriferum* mutants. In *Barley genetics III* (ed. H. Gaul), pp. 20–22. Munich: Karl Thiemig.

Wettstein-Knowles, P. von 1979 Genetics and biosynthesis of plant epicuticular waxes. In *Advances in the biochemistry and physiology of plant lipids* (ed. L.-Å. Appelqvist & C. Liljenberg), pp. 1–26. Elsevier/North-Holland Biomedical Press.

Unstable genotypes

By A. Durrant

Department of Agricultural Botany, University College of Wales, Penglais, Aberystwyth, Dyfed SY23 3DD, U.K.

Unstable genotypes are normally recognized in cultivated plants by the appearance of coloured spots or flecking which are easily seen and scarcely affected by the environment. Unstable genes giving variation in characters such as plant mass, height, yield and earliness are unlikely to be recognized because normally it would be virtually impossible to distinguish such variation from environmental variation and other genetic variation normally ascribed to segregating genes, without deliberate search and detailed analysis. Continuous variation may contain a host of instabilities, or their more stable products, which could have had their origins in the many normal processes of differentiation. It is not known how easily heritable changes can be induced by the environment in plant species in general, but in flax and *Nicotiana rustica* environmentally induced changes giving large relatively stable differences in plant mass, height and flowering time provide a means of studying the behaviour of unstable genotypes affecting these characters. Crosses between environmentally induced flax types and varieties show that the phenotypic differences are not essentially any different from, and could be dispersed unnoticed among, the rest of the genetic and environmental variation. They also show additivity, dominance and gene interaction, and they can be the cause of asymmetric response to selection and inbreeding depression.

1. Major gene differences

Unstable genotypes are well known in cultivated plants. They are due to genetic changes occurring in cells at frequencies well above the frequencies that one normally associates with gene mutations and are confined, as far as one can judge, to a particular chromosome region for any one character showing the particular instability. Genetic changes occurring in unstable genotypes are not due to classical gene mutations, deletions or base changes, but to changes in gene regulation, which may be reversible or maintained for indefinite periods. They are, however, frequently referred to as mutations, and the factors concerned as highly mutable genes.

The presence of an unstable gene is often first revealed by the occurrence of a mosaic of cells, or of colour flecking of a tissue. For example, in *Antirrhinum majus* (Harrison & Fincham 1964) an inactivated anthocyanin gene, $Pal \to pal^{rec}$ gives white flowers instead of red. But pal^{rec} is unstable and is frequently reactivated, $pal^{rec} \to Pal$. Whenever this occurs, red spots appear on the white petals, so that the instability of this allele is easily recognized and the frequency of reactivation assessed. Such changes may or may not be an aberrant form of differentiation, but they are of particular interest because the reactivation, or other change in gene regulation, can be transmitted by the gametes to the next generation and when this occurs new types arise.

The process of inactivation or reactivation is generally held to be due to changes in heterochromatization (see, for example, Hagemann & Snoad 1971), to the movement of controlling elements (see, for example, McClintock 1965; Peterson 1976), or to changes in the number of repeated DNA sequences of some kind (see, for example; Brink et al. 1968; Ritossa 1970), but little is known of what controls or regulates the activating or inactivating process. Furthermore, unstable genotypes frequently show a range of activity, perhaps a stepwise change from one level to another. Paramutation (Brink 1960) is a particular form of instability arising when two different alleles are brought together in the heterozygote. One allele, or chromosome segment, apparently induces a change in the activity of, or paramutates, its homologue, a change that may persist, or eventually disappear, in later generations.

Most studies on unstable genotypes have been on the instability of major genes, i.e. genes giving phenotypic differences that are easily seen and which are almost unaffected by the environment, which is of course why they were detected in the first place. Table 1 gives examples of characters scored in *Antirrhinum majus* (Harrison & Fincham 1964), maize (Peterson 1976; Brink 1960), soybean (Peterson & Weber 1969) and tomato (Hagemann & Snoad 1971). Not only are the phenotypes clear cut, but once recognized as being due to unstable genes their stepwise, or range of, activity can often be measured.

TABLE 1. CHARACTERS SCORED IN THE DETECTION AND ANALYSIS OF UNSTABLE GENOPTYES IN SOME CULTIVATED PLANTS

	locus	character
Antirrhinum	*Pal*	red spots on white petals
maize	A_2–E_n	coloured spots on seeds
maize	*R*	mottled pigmentation on seeds
soybean	Y^m	yellow patches on leaves
tomato	*Sulf*	yellow leaves

2. Continuous variation

Unstable genes can also affect continuously varying characters such as plant mass, height, yield and earliness, characters of central importance to plant breeders, although there is no direct evidence that they or their more stable products are widespread. But nor is there evidence that they are not. Unstable genes giving variation in these characters are unlikely to be recognized because normally it would be virtually impossible to distinguish their variation from environmental variation and other genetic variation normally ascribed to the segregation of genes without deliberate search and detailed analysis. Observed continuous variation may be partly due to a multitude of genetic instabilities, or their more stable products, which could have their origin in the many processes of differentiation which are part of the normal growth and development of the individual. Paramutation could easily supply a pseudo-Mendelian framework for heterozygous but homogeneous F_1 plants with release of variation due to gene instability, rather than due to segregation, in the F_2 and later generations. At issue is how widespread these instabilities are and, in the present context, how important they are in plant breeding.

An initial problem is the separation of variation due to unstable genes from that due to the environment, and from continuous genetic variation of the kind considered to be based on Mendelian inheritance. In some cases it is possible that they may be picked up by biometrical methods. Another way is to start with homozygous and homogeneous plants and induce heritable changes in gene regulation by growing the plants in different environments, as has been done with flax (Durrant 1962, 1971) and *Nicotiana rustica* (Hill 1965). These changes need not be regarded as exceptional because for example in *Antirrhinum* (Harrison & Fincham

1964) and in maize (Peterson 1958) the instability of major genes is markedly influenced by the environment, particularly temperature. Unstable genes have also been induced by chemical mutagens. J. Begum obtained unstable major genes in flax by treating the seed with ethylmethanesulphonate, and at least one other mutation affecting plant mass is probably due to a regulatory change of some kind. Paramutants at the R locus in maize are highly sensitive to chemical mutagens (Axtell & Brink 1967).

In so far as the products of unstable genotypes affecting yield can be stabilized and maintained indefinitely, they are of potential interest to the plant breeder, but we would want to know more about their behaviour on crossing. Here we can turn to some crosses made with flax plants in which inherited changes have been environmentally induced.

3. Environmentally induced changes in flax

Relatively permanent heritable changes have been induced in two varieties of flax by growing the plants with additional nutrients, mainly different fertilizer combinations, and at different temperatures. Because heritable changes are environmentally induced in them, the two varieties are called plastic varieties. Table 2 shows some of the characteristics of two types of plants, the large genotroph (L) and the small genotroph (S), induced in the plastic variety Stormont Cirrus (Durrant 1962, 1971; Evans *et al.* 1966; Durrant & Nicholas 1970; Timmis & Ingle 1973; Cullis 1976). The L and S genotrophs breed true in most environments, like two genetically distinct types, and only in some circumstances do they reveal the nature of their origin. For example, when L and S are crossed, thus bringing together their two different levels of gene regulation, the F_1 is genetically unstable, i.e. an overtly unstable genotype has been generated by crossing, with release of genetic variation.

Table 2. Large and small genotrophs induced from the flax variety Stormont Cirrus

(Mean values over years; DNA in arbitrary units.)

	large (L)	small (S)
plant weight/g	47	13
plant height/cm	87	68
total nuclear DNA/pg	115	100
number of rRNA genes	2914	1801
capsules	hairless	hairy

It has been convenient in the past to recognize two classes of genotrophs, the plastic type, like the original Stormont Cirrus plants in which heritable changes can be induced by the environment, and the stable type, like the L and S genotrophs, in which heritable changes are not subsequently easily induced. But these are relative terms, because plants of the Stormont Cirrus variety appear to have gradually lost their plasticity in some characters over some ten generations in the environments in which they have been grown, although phenotypically they do not appear to be different. On the other hand, L and S genotrophs possess a certain degree of plasticity. For example, L plants are normally maintained by growing them each year in a greenhouse for at least the first 5 weeks from sowing. If, instead, the plants are grown out of doors each generation (figure 1) the amount of nuclear DNA drops each year (Durrant & Jones 1971; Joarder *et al.* 1975), until after five generations out of doors they have lost about 15% of their total nuclear DNA, at which point they contain about the same amount of DNA as the small genotroph, S. During this period there is no change in the L phenotype, but in the sixth generation the L plants are liable to change suddenly to approximately the S phenotype. It is reasonable to suppose that other, undetected, changes occur in the genetic material over the years, which could eventually have important effects on the phenotype. The gradual loss of plasticity of Stormont Cirrus plants in some characters over several generations may be due to similar causes.

Crosses of L and S respectively to Stormont Cirrus variety (SC) from which they were induced, give approximately additive effects. Measurements on the fourth generation by Y. Al-Saheal in figure 2a show that the difference in DNA between the two sets of crosses, 4.8 units, is about half the difference between the L and S parents, as would be expected if the contributions of both parents in each case, L and SC, S and SC, were strictly additive. The DNA amounts in the crosses have, however, dropped a little from the mid-parent values, which may be due to the environment or to a gradual shift towards the lower parent in each case. Plant mass and height in these crosses behave similarly.

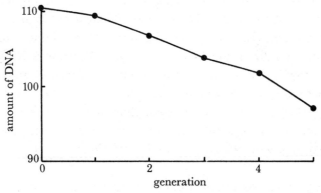

FIGURE 1. Decrease in amount of DNA (arbitrary units) over five generations in the L genotroph of the flax variety Stormont Cirrus when grown out of doors.

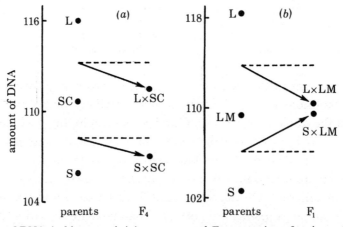

FIGURE 2. (a) Amounts of DNA (arbitrary units) in parents and F_4 generation of reciprocal crosses between L and Stormont Cirrus (L × SC), and S and Stormont Cirrus (S × SC). Broken lines show mid-parent values. (b) Amounts of DNA in parents and F_1 generation of reciprocal crosses between L and Liral Monarch (L × LM), and S and Liral Monarch (S × LM).

4. Outcrosses with the flax genotrophs

Crosses of L and S with other varieties appear to fall into two groups. Either the other variety appears to revert, or cancel out, the heritable changes that originally occurred in the induction of L and S, or the other variety itself could have heritable changes induced in it by L and S (Durrant 1972).

The flax variety Liral Monarch (LM) has about the same, intermediate amount of DNA as Stormont Cirrus, and about the same, intermediate plant mass. When L and S are reciprocally crossed to it, the difference in amount of DNA between the two sets of crosses virtually disappears. Figure 2b shows that the amounts of DNA, measured by O. I. Joarder, are practically the same in the F_1 of the crosses between L and LM, and S and LM, 103.2 and 102.3 units, as in Liral Monarch itself, or in Stormont Cirrus. If the amounts were additive, contributions of

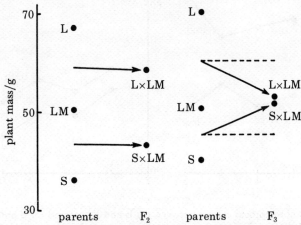

FIGURE 3. Plant masses of parents and F_2 and F_3 generations of reciprocal crosses between L and Liral Monarch (L × LM), and S and Liral Monarch (S × LM) grown in 1971.

TABLE 3. AMOUNTS OF APPARENT REVERSION IN RECIPROCAL CROSSES BETWEEN L AND LIRAL MONARCH, AND S AND LIRAL MONARCH IN DIFFERENT GENERATIONS AND YEARS

	1970	1971	1972
F_1	—	—	complete
F_2	partial	none	complete
F_3	—	complete	complete
F_4	—	—	complete

the parents to the F_1 values would be 106.4 and 99.3 units respectively. Reversion also appears to occur in plant mass but it is often progressive rather than immediate. For example, in figure 3, from work by O. I. Joarder, there is no reversion in an F_2 grown in 1971, but apparently complete reversion in an F_3 generation grown in the same year. The pattern over the years in table 3 shows that the plants must be grown in the right environment for reversion to occur. Evidently 1972 provided the right environment for reversion. Viewed in another way, crossing L or S with Liral Monarch generates plastic plants in which changes in plant mass can be induced in predetermined directions when they are grown in the right environments. L and S crosses with some other flax varieties have roughly similar plant mass patterns to Liral Monarch, whereas L and S crosses to the flax variety Hollandia tend to diverge in plant mass.

An important question in the present context is whether increased plant mass, or yield, can be obtained by using a plant such as the large genotroph L in a crossing programme. Linseed varieties have larger plant masses and seed yields than flax and normally they would not be crossed with flax types for improvement in seed yield unless it were to introduce some particular characteristic. The linseed variety Royal, for example, is about three times larger than Stormont Cirrus. The L and S genotrophs were, however, both reciprocally crossed to Royal and selection for high and low plant mass made over several years. The plant masses of the selection lines in figure 4 have been adjusted against the overall mean plant mass of the three parents, L, S and Royal, in each generation, and in figure 5 against Royal alone for comparison with it.

In the crosses between L and Royal, the plants respond well to high and low selection but they do not exceed the mass of Royal (figure 5). In fact, compared with Royal the high selection line responds once, in F_3, and thereafter stays fairly constant, practically the same as

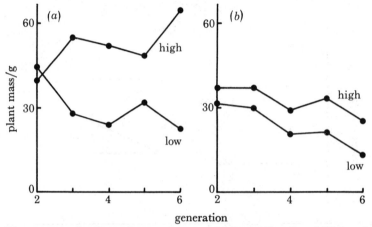

FIGURE 4. Plant masses of reciprocal crosses between (a) L and Royal and (b) S and Royal in high and low selection lines, adjusted against the standardized mean plant masses of the three parents.

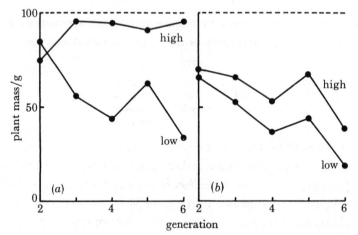

FIGURE 5. Plant masses of reciprocal crosses between (a) L and Royal and (b) S and Royal in high and low selection lines, adjusted against the standardized mean plant masses of Royal only (broken line).

Royal, as though a switch to the higher plant mass had occurred in that generation. It is possible the Royal plant mass might be exceeded were derivatives of crosses between L and S used instead of L, and F_3 backcrossed to Royal or recurrent selection employed. In the crosses between S and Royal there is a grossly asymmetric response to selection. Whether one selects for high or low plant mass there is an inevitable decline. This may be because the S genotroph introduces factors that induce changes in gene regulation in the same direction as that in S, or because the level of gene regulation in S interacts unfavourably with the regulation in Royal. The environments in which they were grown over the years were either unable to switch the level of regulation, or were necessary for the decline to occur. This trend has the appearance and characteristics of inbreeding depression, arising from selfing the reciprocal crosses between S and Royal, but it is due to a level of regulation in S different from that in L.

Although these experiments give some insight into the behaviour of products of unstable gentoypes, they are too meagre to justify general conclusions as to their potential contribution to plant breeding, but some general remarks can be made, which are probably not in disagreement with the results of analyses on the induced changes in *Nicotiana rustica* (Perkins *et al.* 1971; Towey & Jinks 1976). First, the *Linum* crosses probably reveal the hidden behaviour of unrecorded unstable genotypes having similar or more moderate effects on plant size and yield. Secondly, phenotypic differences between relatively stable products of unstable genotypes are not essentially different from any other differences between plants, genetic or environmental, and they could be dispersed unnoticed among the rest of the genetic and environmental variation to which they could make major contributions. Unstable genotypes supplying additional variation can be generated by crosses between the more stable products of unstable genotypes. Thirdly, in so far as unstable genotypes, unrecognized, have contributed to plant improvement in the past, their conscious induction and manipulation could be advantageous in the future. Fourthly, newly introduced, unselected material from natural habitats may respond to the environments into which they are brought in a more direct way than that normally attributed to selection.

References (Durrant)

Axtell, J. D. & Brink, R. A. 1967 Chemically induced paramutation at the *R* locus in maize. *Proc. natn. Acad. Sci. U.S.A.* **58**, 181–187.

Brink, R. A. 1960 Paramutation and chromosome organisation. *Q. Rev. Biol.* **35**, 120–137.

Brink, R. A., Styles, E. D. & Axtell, J. D. 1968 Paramutation: directed genetic change. *Science, N.Y.* **159**, 161–170.

Cullis, C. A. 1976 Environmentally induced changes in ribosomal RNA cistron number in flax. *Heredity, Lond.* **36**, 73–79.

Durrant, A. 1962 The environmental induction of heritable change in *Linum*. *Heredity, Lond.* **17**, 27–61.

Durrant, A. 1971 The induction and growth of flax genotrophs. *Heredity, Lond.* **25**, 513–527.

Durrant, A. 1972 Studies on reversion of induced weight changes in flax by outcrossing. *Heredity, Lond.* **29**, 71–81.

Durrant, A. & Jones, T. W. A. 1971 Reversion of induced changes in amount of nuclear DNA in *Linum*. *Heredity, Lond.* **27**, 431–439.

Durrant, A. & Nicholas, D. B. 1970 An unstable gene in flax. *Heredity, Lond.* **25**, 513–527.

Evans, G. M., Durrant, A. & Rees, H. 1966 Associated nuclear changes in the induction of flax genotrophs. *Nature, Lond.* **212**, 697–699.

Hagemann, R. & Snoad, B. 1971 Paramutation at the *Sulf* locus of *Lycopersicon esculentum*. *Heredity, Lond.* **27**, 409–418.

Harrison, B. J. & Fincham, J. R. S. 1964 Instability at the *Pal* locus in *Antirrhinum majus*. *Heredity, Lond.* **19**, 237–258.

Hill, J. 1965 Environmental induction of heritable changes in *Nicotiana rustica*. *Nature, Lond.* **207**, 732–734.

Joarder, I. O., Al-Saheal, Y., Begum, J. & Durrant, A. 1975 Environments inducing changes in amount of DNA in flax. *Heredity, Lond.* **34**, 247–253.

McClintock, B. 1965 The control of gene action in maize. *Brookhaven Symp. Biol.* **18**, 164–182.

Perkins, J. M., Eglington, E. G. & Jinks, J. L. 1971 The nature of permanently induced changes in *Nicotiana rustica*. *Heredity, Lond.* **27**, 441–457.

Peterson, P. A. 1958 The effect of temperature on the mutation rate of a mutable locus in maize. *J. Hered.* **49**, 120–124.

Peterson, P. A. 1976 Basis for the diversity of states controlling elements in maize. *Molec. gen. Genet.* **149**, 5–21.

Peterson, P. A. & Weber, C. R. 1969 An unstable locus in soybeans. *Theor. appl. Genet.* **39**, 156–162.

Ritossa, F. M. 1968 Unstable redundancy of genes for ribosomal RNA. *Proc. natn. Acad. Sci. U.S.A.* **60**, 509–516.

Timmis, J. N. & Ingle, J. 1973 Environmentally induced changes in RNA gene redundancy. *Nature, new Biol.* **244**, 235–236.

Towey, P. & Jinks, J. L. 1976 The number of phenotypes among the conditioned lines of *Nicotiana rustica*. *Heredity, Lond.* **37**, 357–364.

Discussion

H. REES, F.R.S. (*Department of Agricultural Botany, University College of Wales, Aberystwyth, U.K.*). It is interesting to compare the induced changes in chromosomal DNA in flax with those induced by hybridization in *Nicotiana*. In the hybrid *N. tabacum* × *N. otophora*, certain chromosome segments become amplified to a prodigious degree. In this case the 'conditioning' is a response to a genetic as distinct from an external stimulus. It would be of interest to find out if the molecular changes in the chromosomes of *Nicotiana* are of the same kind as in flax.

A. DURRANT. It is also interesting to note that the amplification occurs in a genus in which heritable changes have been induced by external stimuli. Genetic requirements for environmentally induced heritable changes in amount of DNA have been studied in flax by crossing and backcrossing between a plastic flax variety and a stable linseed variety. The plastic variety apparently possesses a nuclear factor and a cytoplasmic factor that are both lacking in the stable variety, and the nuclear factor to be operative must be transmitted in a nucleus from a plant that possessed the plastic cytoplasm.

C. A. CULLIS (*John Innes Institute, Norwich, U.K.*). The DNA from a number of the genotrophs described by Dr Durrant has been characterized. The ribosomal RNA gene number and a fraction of the intermediately repetitive sequences have been shown to vary between genotrophs. Three intermediately repetitive sequences, which vary in copy number between genotrophs, have been cloned from total flax DNA. These sequences were hybridized to total DNA from a number of genotrophs, which had been digested by the restriction endonuclease *Eco*RI. For one of the cloned sequences, a band of hybridization appeared in the DNA from genotroph L_3 that was not present in the DNA from either L_1 or L_6. The origin of this band, whether or not it is due to transposition, deletion or some other event, is being investigated.

A. DURRANT. The L_3 genotroph has the same phenotype as the L genotroph, but its total amount of DNA has been reduced to the same intermediate amount as in the original variety, by growing the L plants out of doors for three generations. DNA hybridization studies will become increasingly important in establishing the nature of the DNA changes, if not the mechanism of the changes, and one looks forward to seeing these results.

Nuclear instability and its manipulation in plant breeding

By M. D. Bennett

Plant Breeding Institute, Maris Lane, Trumpington, Cambridge CB2 2LQ, U.K.

Nuclear instability occurs spontaneously in a typically very small proportion of cells of every individual, even in crop varieties. Of greatest interest to the cereal breeder are instabilities in the germ line, which produce off-types among progeny, or in the endosperm, which reduce grain quality.

Nuclear instabilities in crop plants merit cytological investigation for several reasons: first, to ensure that biologically possible standards of genetical purity are set for varieties in agriculture; secondly, because once understood, nuclear instability may be usefully applied in plant breeding; thirdly, because nuclear instability is thought to have played a major role in crop plant evolution – understanding the past may help in predicting which new genome combinations will be successful crop species; fourthly, because failure to achieve adequate nuclear stability has played a major role in preventing so many potentially useful plants from becoming crops.

These points are illustrated mainly by reference to three different nuclear instabilities, namely: (1) haploid barley production by genome elimination in some *Hordeum vulgare* × *H. bulbosum* crosses; (2) the action of the *tri* gene in barley to produce about 50% diploid embryo sacs; (3) aberrant endosperm development in hexaploid triticale. Improved seed type in triticale has been achieved by a controlled reduction in rye telomeric heterochromatin. This approach may open the way for a new type of plant breeding, selecting for nucleotypic variation in the amount of non-coding DNA sequences.

Understanding the cellular mechanisms responsible for nuclear stability (or instability) is essential if controlled plant modification based on precise nuclear engineering is to become possible. This understanding can come only from sustained fundamental research.

1. Introduction

For the purpose of this paper, nuclear instability is defined as any deviant nuclear behaviour producing a nucleus (or nuclei) of abnormal structure, karyotype or behaviour, while plant breeding is interpreted in its widest sense to include cytogenetical research that has either already influenced crop variety production or might reasonably be expected to do so in the foreseeable future.

Shortage of space makes it impossible to review here the numerous types of nuclear instability that are known. Instead, this paper presents a few general comments about nuclear instability in the context of plant breeding and then mentions some of the important reasons why it merits sustained experimental attention in crop species and breeder's material. These points are illustrated mainly by reference to three examples of nuclear instability that I have recently studied, which occur at female meiosis or early in seed development. Thus, the present work is restricted to discussing aspects of nuclear instability in those temperate cereal grain crops (notably barley, bread wheat and hexaploid triticale) with which I am most familiar. It is hoped that the phenomena described, and the conclusions drawn, will prove to have a wider relevance.

2. Occurrence of nuclear instability in varieties

Higher plant species have been subject to prolonged natural selection for nuclear stability in their normal environments. Moreover, varieties of crops such as bread wheat and barley, bred for high fertility and uniformity, have also been subjected to intensive breeder's selection for nuclear stability. Nevertheless, nuclear instability occurs spontaneously in a typically very small proportion of cells of every individual, even in such crop varieties. For example, Riley & Kimber (1961) estimated the frequency of aneuploid progeny in four varieties of bread wheat (*Triticum aestivum*) and a variety of cultivated oat (*Avena sativa*) as 1.08% and 1.27%, respectively. These frequencies for hexaploids are typically much higher than in diploids. For example, Sandfaer (1973) estimated the frequency of aneuploid progeny in diploid barley (*Hordeum vulgare*) as only 0.003%. In general, the incidence of nuclear instability increases with the wideness of the cross. Aneuploidy is more frequent in crosses between than within bread wheat varieties.

An increased incidence of nuclear instability may be induced by many genetical and environmental factors including temperature (Bayliss & Riley 1972). Interestingly, nuclear instabilities may be much more frequent in diseased plants. Sandfaer (1973) showed that in 11 spring barley varieties infected with barley stripe mosaic virus, the incidence of aneuploid and triploid plants was respectively about 110 and 48 times higher than in healthy controls.

3. Some reasons why nuclear instability merits attention

Chromosomes and nuclei are self-replicating entities, so that a very low incidence of nuclear instability can have drastic consequences. Nuclear instability can occur in all cell types and throughout the life cycle. However, of greatest importance to the plant breeder are those instabilities that occur at significant frequencies either in the germ line affecting the phenotype of progeny, or in the endosperm and adversely affecting the expression of grain development. For example, aneuploidy (usually caused by pairing failure at meiosis) frequently results in reduced fertility and/or progeny with diverse phenotypes in wheat (Law & Worland 1973; Sears 1954) and barley (Tsuchiya 1967), while mitotic abnormalities in coenocytic endosperm are a frequent cause of sterility and misshapen grain (Moss 1970; Bennett 1974). Thus, nuclear instability merits attention first because it makes achieving the aims set by and for plant breeders (e.g. high fertility and uniformity) more difficult or impossible.

A second important practical reason for studies of nuclear instability concerns the legal and commercial standards of purity set for crop varieties. Riley & Kimber (1961) noted that the frequency of chromosomally variant types produced by nuclear instability in bread wheat is frequently greater than the incidence of morphologically distinguishable 'off-types' tolerated by some seed certification organizations. It is important that high standards of genetic purity are set and maintained in such inbred crops but, in that irregular chromosome constitutions may lead to phenotypic deviations (Sears 1954), it is essential that the required standards of genetic purity are biologically possible.

A general aim of the cytogeneticist is to understand the various causes and forms of nuclear instability, to be able to maximize nuclear stability within crops. In that other scientists are unlikely or unable to study chromosomes and nuclei, the cytologist has a unique responsibility to monitor their behaviour and misbehaviour, and to interpret his observations in relation to the plant breeder's aims and problems. In this connection it is important to remember that in wheat, barley, rye and many other cereals there is only a single female meiocyte, functional megaspore, egg and central cell per floret. In contrast with its occurrence on the male side of two-track heredity, the consequences of nuclear instability in these unique female cells are unavoidable. The basic unit of grain production is the floret and therefore the potential number of grains is the same as the number of fully formed florets. However, this potential is rarely if ever achieved in the field, the shortfall commonly ranging from only about 3–5 % of first and second florets in the best crops of bread wheat varieties (J. Bingham, personal communication) to about 20 % of florets in a two-rowed barley crop (Barling 1979). Nuclear instability in the female germ line has been recognized as one possible cause of sterility (Riley & Kimber 1961) but, as far as I am aware, its incidence has not been carefully quantified cytologically in any cereal grown either under equable conditions (when its importance is minimal) or under environmental stress such as drought, air-frost or high temperatures, when its practical significance may be greater. It is therefore suggested that such studies are long overdue because information on the incidence and causes of nuclear instability in the female germ line may indicate the need, if not the means, for a useful improvement in fertility in some crop plants.

4. Genome elimination

Although nuclear instability is generally undesirable because of its deleterious consequences, this is not invariably so. Indeed, there is one clear example in which spontaneously occurring nuclear instability has recently been manipulated to good effect by cytogeneticists; namely, genome elimination in certain hybrids.

Nuclear instability and chromosome loss is a common but variable phenomenon in interspecific hybrids, but in an extreme form all the chromosomes of one parent species are retained while all those of a second parent species are eliminated from dividing hybrid cells. Such genome elimination occurs in some crosses between diploid *Hordeum vulgare* and diploid *H. bulbosum*, where the *H. bulbosum* genome is eliminated leaving nuclei with a haploid *H. vulgare* complement (Kasha & Kao 1970; Subrahmanyam & Kasha 1973). Although elimination can occur at any time in the life cycle (Humphries 1978), it is frequently completed in young embryo cells soon after fertilization (Bennett *et al.* 1976).

Once the nature of the nuclear instability was understood, its advantage was soon realized. Thus, by using colchicine it is possible to double the chromosome number of haploid barley plants produced from a hybrid by genome elimination to give immediately homozygous plants (dihaploids), thereby removing the need for seven or eight generations of selfing from a barley breeding programme. This possibility has been applied in Cambridge and elsewhere in both spring and winter barley breeding programmes (although the time-saving over conventional breeding methods is greatest in winter barley) to produce thousands of dihaploids from plants in early generations. Two such lines produced at the Plant Breeding Institute, Cambridge, are showing promise and are being bulked before assessment in national trials, while in Canada a spring barley variety 'Mingo' produced by this means has been licensed for growing in Ontario only 5 years after the initial cross (Ho & Jones 1980).

Genome elimination in *Hordeum* is an example of nuclear instability that frustrated the purpose of the original interspecific cross (i.e. to make a hybrid) but was then applied in plant breeding after cytogeneticists had studied its nature, recognized its potential and developed its application. Knowledge of this interesting and useful phenomenon stems from a surprise result from wide-crossing experiments (Davis 1958). Indeed, neither its discovery nor its application were expected or planned. Thus, the recent history of this particular nuclear instability clearly shows how narrow the gap may be between fundamental cytogenetic studies and their practical application in plant breeding.

Similar genome elimination has been demonstrated in some hybrids between bread wheat and tetraploid *H. bulbosum* to yield wheat haploids (Barclay 1975). However, extending its application to wheat breeding has proved difficult because most varieties of bread wheat currently in agriculture possess genes that prevent crossing with *H. bulbosum* (Snape et al. 1979). Consequently, the possibility is being explored of producing dihaploids for use in wheat breeding programmes, either by ensuring that the genes for non-crossability with *H. bulbosum* are absent from future breeders' material, or by identifying clones of *H. bulbosum* less sensitive to their action (J. Snape, personal communication). Meanwhile, genome elimination is recognized as a potentially useful means of producing haploids in high frequencies in many other crops. Consequently, fundamental research is being conducted in several countries, to elucidate the cellular processes responsible.

5. Genome doubling in barley

Another example of nuclear instability in barley that seems to have potential for useful manipulation concerns genome doubling. Ahokas (1977) described a recessive mutant that he named triploid inducer (*tri*) in a single plant selection of the variety 'Paavo'. Diploid plants with 14 chromosomes homozygous for *tri* are fully viable and of normal fertility, but have two types of seed distributed at random within all spikes. One type has a normal endosperm and a diploid embryo, but the other has a thin endosperm and a triploid embryo. It was recently shown (Finch & Bennett 1979) that *tri* acts to produce chromosomally doubled functional megaspores (and subsequently unreduced egg nuclei) by suppressing the second meiotic division in about 50 % of florets. However, *tri* does not similarly affect male meiosis.

Chromosome doubling with colchicine still proves difficult in some materials. It would be useful, therefore, if a gene like *tri*, which controls an exact doubling of the chromosome number, could be used to achieve doubling more easily, more frequently, or with a greater control of its timing. As a first step it was interesting to ask whether *tri* acts at higher ploidy levels. Autotetraploid *tri* plants were prepared by doubling a *tri* diploid with colchicine, and these produced about 50 % hexaploid progeny, showing that *tri* acts at the higher ploidy level (R. A. Finch & M. D. Bennett, unpublished results). Hitherto, autohexaploid barley plants have been so rare as to merit special publication (Rommel 1960). By using *tri*, one hundred plants counted as hexaploids were easily obtained. Another potential use of *tri* is to make new genetic stocks more easily than before. For example, it should be possible to use the abundance of diploid and triploid *tri* progeny to make a new set of trisomics which, when crossed with *Hordeum bulbosum* should (after elimination of *H. bulbosum* chromosomes), yield significant numbers of tetrasomic plants with 16 chromosomes, if these are viable. Hitherto, such plants have been rare in barley (Tsuchiya 1967).

6. Apomictic cereals

Unreduced gametes are known to occur at low frequencies (i.e. < 1 %) in many crop species, and are quite common on both the male and female side in a few sexual species (e.g. *Saccharum officinarum*), but they are most frequent in gametophytic apomicts. The notion of developing apomictic lines of cereal crops has certain attractions (Asker 1979), notably the possibility of fixing heterosis in crops such as hybrid maize or even bread wheat. Attention has mainly been focused on the possibility of introducing the gene complex controlling apomixis from another species by wide crossing. However, the way forward may lie in identifying within the crop species genes controlling the essential steps for apomixis, e.g. regular production of unreduced eggs, and parthenogenesis (the former by screening for high frequencies of triploids, and the latter by screening for high frequencies of haploids). In other words, it may be important to combine genes that fix as a regular feature of female germ line development something that, as an isolated event, would be termed nuclear instability.

7. Nuclear instability and evolution

The action of *tri* in causing genome doubling contrasts in evolutionary potential with that of the genes controlling genome elimination in many *Hordeum* species. If, as seems likely, both genes exist in wild *Hordeum* species then a prospect is opened of genomic 'snakes and ladders', with new polyploids being produced after genome doubling more easily and frequently than has been usually envisaged, and new diploids whose genomes have been modified during their association with other diploid genomes in polyploid species, being produced by genome elimination in hybrids. Genes in diploids that affect chromosome pairing in polyploids, but not apparently in the diploid itself (Dover & Riley 1972), may represent preadaptation for polyploidy, but they may also provide evidence of diploids being derived from polyploids by genome elimination.

Another reason why nuclear instability merits attention concerns the important role that it is believed to have played in crop plant evolution. For example, tetraploid and hexaploid wheats are thought to have arisen from parent diploids by hybridization followed by 'nuclear doubling', i.e. some unknown nuclear instability. Greater understanding of the evolutionary biology of crop genera should indicate the conditions for, and the means of, further or alternative crop improvement. Knowledge of the genetic architecture of bread wheat certainly contributed to the discovery of the *Ph* locus on chromosome 5B of bread wheat, whose activity is critical for maintaining nuclear stability at meiosis. The exciting and extensive manipulation of the wheat genome that this important discovery made possible is described in the accompanying papers by Thomas and Riley *et al.* (both in this symposium).

Realization of the important role of polyploidy in crop evolution resulted in the synthesis of a large number of new polyploids (see, for example, Bell & Sachs 1953), but few have become crops mainly because of nuclear instability at meiosis or in young seeds. For example, there were high hopes for autotetraploid barley (Nilan 1964). Initially its meiotic stability and fertility were poor but both were considerably improved by selection. However, the mechanism(s) responsible for the improvement are unknown. The pragmatic approach failed, and the improvement was just insufficient for its acceptance as a significant crop. Nevertheless, further understanding of the mechanisms of chromosome pairing in stable tetraploid species may suddenly provide a more rational basis for completing the development of many plants like autotetraploid barley as useful crops.

Attempts to construct new allopolyploid cereals have also been largely unsuccessful with the exception of hexaploid triticale, which seems certain to become a significant cereal crop species (Muntzing 1979). Triticale's incipient success shows that the concept of man-made crop species is feasible. However, triticale's development has been pragmatic, and consequently adds nothing to our ability to predict which other novel combinations of genomes can be made stable and productive in agriculture. This ability requires an understanding of the basic mechanisms determining nuclear stability or instability. Thus, an important value of triticale, whose whole development during about the past 30 years is known and open to investigation, is as a test-bed for experiments to establish the basis of its improvement. As an example of this approach, work on the cause and control of nuclear instability in triticale endosperm will be described.

8. Nuclear instability in triticale endosperm

In about 1970 it was clear that development of hexaploid triticale (comprising the chromosome complements of tetraploid *Triticum durum* and diploid *Secale cereale*) faced two major problems, namely sterility and shrivelled grain at maturity. Nuclear instabilities occur infrequently in young endosperms of all cereals, but observations of hexaploid triticale at this stage revealed a particularly extreme and frequent sequence of aberrant nuclear behaviour (Bennett, 1974). In wheat, rye and triticale, early endosperm development is coenocytic until about 1000–2000 nuclei are formed. Moreover, nuclear development within the coenocyte is highly synchronous, and the nuclear doubling time is very short compared with embryo and other cell cycle times (Bennett *et al.* 1975).

Nuclear instability in triticale endosperm often began with the formation of a chromatin bridge at anaphase (Bennett 1974). This would have been broken by cell wall formation in a cellular tissue but it persisted throughout interphase in the endosperm coenocyte. As a result, restitution nuclei were formed whose DNA contents doubled at successive rounds of DNA synthesis in the coenocyte. If the initial bridge occurred early in the coenocytic phase, then monstrous aberrant endosperm nuclei with 128, 256 or 512 times their normal DNA contents resulted, and the endosperm invariably aborted without becoming cellular. However, if the initial bridge formed late during the coenocytic phase, then cells with considerably higher ploidy levels than normal (but lower than those just mentioned) were formed at endosperm cellularization. Such aberrant cells, which were often formed in groups, subsequently either aborted or failed to contribute normally to endosperm cell division. Overcoming this instability would clearly depend on understanding the cause of the initial anaphase bridge.

Rye has about 35% more DNA than the largest diploid genome in tetraploid wheat. Moreover, all seven replicated rye chromosomes contain more DNA (4.4–4.9 pg) than the largest wheat chromosome (3.95 pg) (Gustafson & Bennett 1976). Unlike wheat chromosomes, rye chromosomes have large telomeric segments of heterochromatin on one or both arms that are selectively stained by Giemsa (Gill & Kimber 1974; Bennett *et al.* 1977). These segments account for about 18% of the total rye DNA (Bedbrook *et al.* 1980), are of recent origin, and contain some families of very highly repeated DNA sequences which are not detected in wheat (see also Flavell, this symposium). This and the fact that development is unimpaired or even improved after their deletion suggests that they are non-coding DNA. Lima-de-Faria & Jaworska

(1972) showed that at DNA synthesis phase(s) in rye the telomeric segments are the last to replicate, and that the last 33% of S-phase involves synthesis of telomeric segments alone.

The significance of these interspecific differences in DNA amount and type is that both an increased DNA amount per diploid genome and an increased amount of late-replicating heterochromatin generally result in slower cell development. Moreover, rye and wheat are known to conform to this expectation. Thus, when rye and wheat are compared under controlled conditions, rye has a longer meiosis, pollen development time, cell cycle time in embryo

TABLE 1. THE PERCENTAGE OF ABERRANT NUCLEI IN 3 DAYS OLD COENOCYTIC ENDOSPERMS OF THE SEVEN DISOMIC ADDITION LINES OF HOLDFAST–KING II WITH AND WITHOUT DELETIONS OF TELOMERIC HETEROCHROMATIN

(From Bennett (1977).)

Addition	With deletion	Without deletion
I (5R)	—	0.51
II (6R)	—	1.30
III (2R)	0.28	—
IV (4R/7R)	—	2.75
V (1R)	—	1.05
VI (3R)	0.13	—
VII (7R/4R)	0.20	—
mean	0.20	1.40

cells, nuclear doubling time in coenocytic endosperm, and S-phase in root-tip cells. Graves (1972) has shown that in hamster–mouse somatic cell hybrids, DNA synthesis begins synchronously in both complements, but the different durations and intragenomic patterns of synthesis in the two parent genomes are retained and appear to be regulated autonomously. If a similar situation applied in triticale, and the replication of the rye genome takes longer to complete S than wheat, then under certain circumstances rye chromosomes may enter mitosis (or meiosis) having failed to complete replication at one or both telomeres, resulting in bridge formation at anaphase.

The above observations and facts led to the development of a model, first published in 1974, for a causal chain linking rye heterochromatin with shrivelled grain in triticale. Thus, it was proposed that (1) late-replicating DNA (mainly telomeric heterochromatin) in rye chromosomes causes bridge formation at anaphase; (2) such bridges cause the production of abnormally polyploid endosperm nuclei; (3) that such aberrant nuclei cause either sterility (if they exceed a certain DNA C-value) or shrivelled grain (if they occur in significant numbers with lower DNA C-values). Thus, bridge formation causes either sterility or shrivelled grain depending on its timing during the coenocytic endosperm deveopment (Bennett 1974).

Evidence strongly supporting each step in this causal sequence has been obtained (Bennett 1977). First, C-banding of triticale endosperm mitoses has shown that anaphase bridges are invariably caused by rye chromosomes. Moreover, such bridges are usually caused by non-separation of telomers (the last-replicating segment). Secondly, further observations of triticale endosperm have repeatedly confirmed that persistent bridges develop into aberrant endosperm nuclei in disomic addition lines of King II rye to Holdfast wheat with or without deletions involving the loss of telomeric heterochromatin at one telomere. There was a significantly higher proportion ($p = < 0.05$) of aberrant nuclei in the four addition lines without the deletion of telomeric heterochromatin than in the three lines with such deletions (table 1). Thirdly, close positive correlations have been shown between the incidence of aberrant nuclei in coenocytic endosperm and the degree of shrivelling at maturity. The more aberrant nuclei there are during early endosperm development, the more shrivelled is the grain at maturity.

By 1976 the evidence that the presence of later-replicating DNA at rye telomeres was a major cause of grain shrivelling in many triticales was convincing. This provoked the question of what could be done to stabilize nuclear development at the critical stage of early endosperm development. It was decided to try to arrange for the absence of the rye telomeric heterochromatin in such plants. Four possible methods were considered.

1. To screen the genus *Secale* for taxa with less DNA and less telomeric heterochromatin than cultivated rye that could be used as parents for new primary triticales. Such taxa exist (Bennett *et al.* 1977).

2. To develop wheat–rye substitution lines. In about 1974 it became clear that most improved triticales from CIMMYT contained less than 14 rye chromosomes owing to their having wheat–rye substitutions (Merker 1975).

3. To utilize rye chromosomes occasionally found in triticale modified by the loss of some or all of the heterochromatin at one or both telomeres (Bennett 1977).

4. To develop a treatment that selectively removes telomeric segments from rye chromosomes in triticale.

In collaboration with J. P. Gustafson (University of Manitoba), I elected to explore the third possibility. It was therefore decided to screen hexaploid triticale lines for modified rye chromosomes with reduced or no heterochromatin at telomeres, to combine these in a common triticale background, and to study the effect on grain development.

TABLE 2. COMPARISON OF THE PERCENTAGE OF ABERRANT NUCLEI IN 20 ENDOSPERMS FIXED 72 h AFTER POLLINATION FOR PLANTS GROWN AT 20 °C, AND THE YIELD, TEST MASS AND 1000 KERNEL MASS IN FIELD-GROWN PLANTS FOR TRITICALE LINES DIFFERING FOR THE PRESENCE ($+$) OR ABSENCE ($-$) OF HETEROCHROMATIN ON THE SHORT ARM (S) OR THE LONG ARM (L) OF RYE CHROMOSOME 7, OR 6 AND 4.

(The number of location test years are given in parentheses.)

hexaploid triticale line	rye chromosome heterochromatin comparison	aberrant endosperm nuclei (%)	yield† t/ha	test mass† kg/hl	1000 kernel mass†/g
DR-IRA HH	7L+	5.2	4.79 (10)	65 (10)	45 (10)
DR-IRA EE	7L−	4.0	5.19*	64*	49**
1204	6S+ 4S+	6.8	2.47 (3)	62 (5)	48 (5)
1206	6S− 4S+	3.8	3.48**	63**	50**
1208	6S− 4S−	2.0	3.01	67	55

* $p = 5\%$; ** $p = 1\%$.
† Unpublished data, courtesy of Dr J. P. Gustafson, University of Manitoba.

Two rye chromosomes (4R and 6R) modified by the loss of heterochromatin in their short arms, were found by Gustafson in the line Miscellaneous 3636 and the variety Rosner, respectively. In addition, Merker (1976) described isogenic lines for the presence (HH) and absence (EE) of the largest telomeric C-band in the complement on rye chromosome 7 in the hexaploid triticale DR-IRA. Comparison of early endosperm development in lines homozygous for one or two of these chromosomes with or without telomeric heterochromatin showed, as expected, that the proportion of aberrant endosperm nuclei at the coenocytic stage was reduced when rye telomeric heterochromatin was removed (table 2). Moreover, the effect was additive when two modified rye chromosomes were combined in line 1208 (figure 1). While these and similar results for other lines (M. D. Bennett & J. P. Gustafson, unpublished) are very encouraging, the crucial test of the value of the modified rye genome in triticale must be its performance in adequately replicated field trials. The first results from such tests (table 2) are also very encouraging. For example, both the yield and the mean kernel mass were about 8% higher in DR–IRA without heterochromatin on the long arm of 7R than in DR–IRA with such heterochromatin. Similarly, in line 1206 (with modified 6R) and 1208 (with modified 4R and 6R) mean kernel mass was respectively 4 and 14.5% higher than in line 1204 (with unmodified 4R and 6R). Moreover, test mass (which is often inversely proportional to shrivelling) showed a significant increase of 8% with reduced telomeric heterochromatin in these three lines.

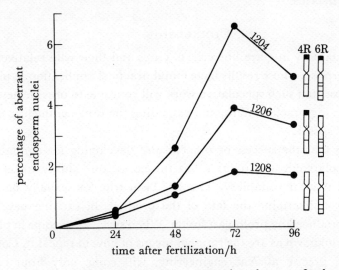

FIGURE 1. The mean percentage of aberrant nuclei in coenocytic endosperms fixed up to 96 h after pollination in triticale 1204, 1206 and 1208 lines with or without telomeric heterochromatin on the short arms of rye chromosomes 6 and 4.

Taken together, the available results seem to confirm that the presence of rye telomeric heterochromatin can have a major deleterious effect in many triticales on several interrelated grain characters of prime agronomic importance including shrivelling, kernel mass, fertility and yield. Consequently, work is continuing to produce triticales in which all of the rye chromosomes are modified by the loss of most, if not all, of their late-replicating heterochromatin. This work is funded in Cambridge by the Overseas Development Administration, and in Winnipeg by CIMMYT. Examples of five of the seven rye chromosomes, and seven of the eleven telomeres, have so far been identified in a modified condition. If they were all combined in a single line the normal appearance of the C-banded rye complement would be significantly altered, and would involve an estimated reduction in the DNA content of the rye genome in triticale of about 10%.

9. A NEW DIMENSION IN PLANT BREEDING

It seems worthwhile to emphasize the novel features of the chromosome engineering being attempted in triticale. First, to improve its agronomic character it is intended to select for a controlled reduction in the mass of DNA in the rye genome in triticale of about 18%. There seems little doubt that this is feasible. Progress so far, involving three telomeres, has already involved an estimated reduction in the mass of rye DNA of about 4–5%. Secondly, most, if not all, of the DNA to be lost is non-genic, and the desired change does not involve selection for or changes in any coding sequences, but only for an improved nucleotype. Thirdly, if this approach results in an improved agronomic performance in triticale, a new dimension will have been added to plant breeding, since similar nucleotypic manipulation is potentially applicable to other interspecific hybrids whose parent species display nucleotypic differences similar to those between wheat and rye. Examples may include *Zea* × *Sorghum* and various interspecific pulse crop hybrids.

10. CONCLUSION

Studies of nuclear instability in germ line cells of crops and their wild relatives have been a source of interest and surprise, whose results have found practical application in plant breeding. It is reasonable to suppose that such speculative work will continue to provide new or improved means for achieving existing practical aims, and extending the range of nuclear manipulations at the plant breeder's disposal.

It is worth repeating that the success or otherwise of developing for agriculture so many apparently attractive polyploids and hybrids has turned on our ability to select lines with acceptably low rates of nuclear instability. This has been true for sexually derived material and no doubt it will yet determine the fate of the products of exciting new technologies, including new hybrids produced by protoplast fusion. When success has come in plant breeding, the reason why is often unknown as are the rational means of how to repeat it. Controlled plant modification, based on precise nuclear engineering, will come only from concerted and sustained fundamental research leading to a deeper understanding of basic cellular mechanisms, including those that determine nuclear stability and instabilities.

REFERENCES

Ahokas, H. 1977 A mutant of barley: triploid inducer. *Barley Genet. Newsl.* **7**, 4–6.

Asker, S. 1979 Progress in apomixis research. *Hereditas* **91**, 231–240.

Barclay, I. R. 1975 High frequencies of haploid production in wheat (*Triticum aestivum*) by chromosome elimination. *Nature, Lond.* **256**, 410–411.

Barling, D. M. 1979 *Winter barley study*. Cirencester: Royal Agricultural College.

Bayliss, M. W. & Riley, R. 1972 An analysis of temperature-dependent asynapsis in *Triticum aestivum*. *Genet. Res.* **20**, 193–200.

Bedbrook, J. R., Jones, J., O'Dell, M., Thompson, R. D. & Flavell, R. B. 1980 A molecular description of telomeric heterochromatin in *Secale* species. *Cell* **19**, 545–560.

Bell, G. D. H. & Sachs, L. 1953 Investigations in the Triticinae. II. The cytology and fertility of intergeneric and interspecific F₁ hybrids and their derived amphidiploids. *J. agric. Sci., Camb.* **43**, 104–115.

Bennett, M. D. 1974 Meiotic, gametophytic and early endosperm development in Triticale. In *Triticale* (ed. R. MacIntyre & M. Campbell), pp. 137–148. I.D.R.C.

Bennett, M. D. 1977 Heterochromatin, aberrant endosperm nuclei and grain shrivelling in wheat-rye genotypes. *Heredity, Lond.* **39**, 411–419.

Bennett, M. D., Finch, R. A. & Barclay, I. R. 1976 The time rate and mechanism of chromosome elimination in *Hordeum* hybrids. *Chromosoma* **54**, 175–200.

Bennett, M. D., Gustafson, J. P. & Smith, J. B. 1977 Variation in nuclear DNA in the genus *Secale*. *Chromosoma* **61**, 149–176.

Bennett, M. D., Smith, J. B. & Barclay, I. R. 1975 Early seed development in the Triticaeae. *Phil. Trans. R. Soc. Lond.* B **272**, 199–227.

Davies, D. R. 1958 Male parthenogenesis in barley. *Heredity, Lond.* **12**, 493–498.

Dover, G. A. & Riley, R. 1972 Variation at two loci affecting homoeologous meiotic chromosome pairing in *Triticum aestivum* × *Aegilops mutica* hybrids. *Nature, new Biol.* **235**, 61–62.

Finch, R. A. & Bennett, M. D. 1979 Action of triploid inducer (*tri*) on meiosis in barley (*Hordeum vulgare* L.). *Heredity, Lond.* **43**, 87–93.

Gill, B. S. & Kimber, G. 1974 The giemsa C-banded karyotype of rye. *Proc. natn. Acad. Sci. U.S.A.* **71**, 1247–1249.

Graves, J. A. M. 1972 Cell cycles and chromosomes replication patterns in interspecific somatic hybrids. *Expl Cell Res.* **73**, 81–94.

Gustafson, J. P. & Bennett, M. D. 1976 Preferential selection for wheat–rye substitutions in 42-chromosome Triticale. *Crop. Sci.* **16**, 688–693.

Ho, K. M. & Jones, G. E. 1980 Mingo barley. *Can. J. Genet. Cytol.* **60**, 279–280.

Humphreys, M. W. 1978 Chromosome instability in *Hordeum vulgare* × *H. bulbosum*. *Chromosoma* **65**, 301–307.

Kasha, K. J. & Kao, K. N. 1970 High frequency haploid production in barley (*Hordeum vulgare* L.). *Nature, Lond.* **225**, 874–876.

Law, C. N. & Worland, A. J. 1973 Aneuploidy in wheat and its uses in genetic analysis. *Rep. Pl. Breed. Inst. 1972*, pp. 25–65.

Lima-de-Faria, A. & Jaworska, H. 1972 The relationship between chromosome size gradient and the sequence of DNA replication in rye. *Hereditas* **70**, 39–58.

Merker, A. 1975 Chromosome composition of hexaploid triticale. *Hereditas* **80**, 41–52.

Merker, A. 1976 The cytogenetic effect of heterochromatin in hexaploid triticale. *Hereditas* **83**, 215–222.

Moss, J. P. 1970 Endosperm failure and incompatibility in crosses between *Triticum* and *Secale*. *Chromosomes Today* **3**, 124–132.

Muntzing, A. 1979 Triticale, results and problems. *Fortschr. PflZucht. Beih. Z. PflZucht.*, no. 10. (103 pages.)

Nilan, R. A. 1964 The cytology and genetics of barley, 1951–1962. *Res. Stud. Wash. State. Univ.* **32** (1), monogr. Suppl. no. 3. (278 pages.)

Riley, R. & Kimber, G. 1961 Aneuploids and the cytogenetic structure of wheat varietal populations. *Heredity, Lond.* **16**, 275–290.

Rommel, M. 1960 The occurrence of euploid and aneuploid hexaploid plants within the offspring of artificially induced tetraploid *Hordeum vulgare* L. *Can. J. Genet. Cytol.* **2**, 199–200.

Sandfaer, J. 1973 Barley stripe mosaic virus and the frequency of triploids and aneuploids in barley. *Genetics, Princeton* **73**, 597–603.

Sears, E. R. 1954 The aneuploids of common wheat. *Res. Bull. Mo. agric. Exp. Stn* no. 572. (59 pages.)

Snape, J. W., Chapman, V. C., Moss, J., Blanchard, C. E. & Miller, T. E. 1979 The crossabilities of wheat varieties with *Hordeum bulbosum*. *Heredity, Lond.* **42**, 291–298.

Subrahmanyam, N. C. & Kasha, K. J. 1973 Selective chromosomal elimination during haploid formation in barley following interspecific hybridization. *Chromosoma* **42**, 111–125.

Tsuchiya, T. 1967 The establishment of a trisomic series in a two-rowed cultivated variety of barley. *Can. J. Genet. Cytol.* **9**, 667–682.

# Interspecies hybrids and polyploidy

By E. L. Breese†, E. J. Lewis† and G. M. Evans‡

† *Welsh Plant Breeding Station, Plas Gogerddan,*
Aberystwyth, Dyfed, SY23 3EB, U.K.
‡ *Department of Agricultural Botany, University College of Wales,*
Aberystwyth, Dyfed, SY23 3DD, U.K.

Polyploidy has featured strongly in plant evolution as a means of conserving favoured hybrid combinations during sexual reproduction. Combining different genomes is likely to extend adaptation rather than increase yield *per se*. Subsequent stabilization by polyploidy depends on the degree of differentiation of the genomes and on the breeding system, but success is mediated by low chromosome numbers while reduced fertilities may be offset where the crop is harvested primarily for vegetative parts and/or is perennial.

In old established grain crops like the cereals, raw synthetic polyploids are not likely to offer immediate advantages but may be subsequently improved by arduous selection in an extended gene pool (e.g. triticale). The relatively undeveloped herbage grasses offer unique opportunities. Flexibility in grass swards is presently sought through unstable mixtures of races and species. It is the breeders' aim to combine genetically the complementary features of these in stable varieties. Agronomically useful hybrids between diploid *Lolium* species have been stabilized at the tetraploid level through tetrasomic inheritance reinforced by a degree of preferential pairing. This preferential pairing may be genetically enhanced, thus raising the possibility of producing a new agriculturally useful amphiploid ryegrass species. Prospects for developing useful amphiploid hybrids between less closely related *Lolium/Festuca* species is considered and related to more limited objectives of transferring desirable genes or gene complexes.

Introduction

Induced polyploidy has two major consequences. The first is in the development of the plant, chiefly through an increased cell size and related pleiotropic effects including larger but fewer plant parts (the gigas effect). The second is on the genetic system, particularly in preserving hybrid combinations over sexual generations through genome duplication. Almost without exception it is the latter property that has featured in natural plant evolution in the development of race and species hybrids with wider adaptability and a greater pioneering potential than their parents (Stebbins 1950). Lewis (1967) developed this theme to point out that there is little evidence that natural polyploids display what he terms exophenotypic effects (= gigas characteristics) and considers that the diploid size must be retained or regained during evolution. He further points out that since plants can, and usually do, achieve polyploidy in certain somatic cells through controlled endoreplication, the significance of polyploidy in evolution must be in the qualitative rather than quantitative aspects of the gene complement. Schwanitz (1953) also points to the fact that many 'gigas' characteristics associated with polyploidy can be achieved through selection at the diploid level. It is therefore logical to consider that the greatest plant breeding benefits will derive from the use of polyploidy in developing race and species hybrids (allopolyploidy) rather than from autopolyploidy *per se*.

Application of polyploidy to plant breeding

The discovery of the chromosome-doubling properties of colchicine more than 40 years ago raised great expectations of the use of induced polyploidy in genetic manipulation. Its use is attempted in three ways: (i) induced autopolyploidy in an attempt to improve yield or quality (or appearance in the case of ornamentals) by capitalizing on the developmental aspects of large cells and large parts; (ii) to restore fertility to sterile species hybrids with a view to stabilizing them genetically (amphiploidy or allopolyploidy); (iii) to serve as a genetic bridge between species in the transfer of gene complexes (Stebbins 1956; Dewey 1980).

A number of factors predispose success. First, the two near-universal consequences of induced polyploidy are increased cell size and decreased fertility (Eigsti & Dustin 1955). Thus crops that are harvested chiefly for their vegetative parts may enjoy advantages that are less likely to be outweighed by reduced fertilities than seed crops. This advantage also applies to crops which are perennial and/or asexually propagated (e.g. potatoes, herbage legumes and grasses).

Secondly, experience has shown the importance of chromosome number and present ploidy state. Many species have achieved optimal numbers beyond which growth and development is inhibited. Similarly, most taxa have an optimal ploidy level and further additions produce disharmonious growth or are difficult to stabilize. Amphiploid hybrids above this critical level often lose chromosomes or whole genomes in later generations. In many crop plants this optimal ploidy level appears to be hexaploid (Stebbins 1950). Thus for instance, in triticale, breeding has been more successful with tetraploid wheats ($2n = 28$), which give hexaploid ($2n = 42$) hybrids in crosses with diploid rye ($2n = 14$) rather than with hexaploid bread wheat (*Triticum aestivum*) ($2n = 42$), which yield octoploid ($2n = 56$) hybrids (Larter 1976). Other examples are frequent in the literature and some are cited by Dewey (1980). The conclusion is that beyond a certain ploidy level we may be forced to seek improvement through the controlled introgression of genes or gene complexes rather than through genome incorporation.

Thirdly, the breeding system is important. Experience has shown that in polyploid breeding it is important to assemble a large number of raw polyploids that can be intercrossed to form a large gene pool followed by selection for improved performance. This is true for autopolyploid and allopolyploid breeding, and success in most polyploid breeding programmes has undoubtedly been achieved by these means. Cross-fertilizing species are at an advantage in securing this extended gene pool but there are important differences in the consequences of inbreeding and outbreeding with respect to polyploidy which will emerge later.

Principles determining breeding strategies in allopolyploid breeding

Principles that govern our approach to a breeding programme are of two sorts, agronomic and genetic.

From the agronomic standpoint it has to be borne in mind that wide hybrids and their allopolyploid derivatives are usually successful in colonizing new habitats distinctly different from those to which their progenitors have become adapted. The more radical the change in genetic constitution the more radical is the change required in the environment (Stebbins 1956; Tigerstedt 1978). For well adapted, highly selected and intensively managed crops like the cereals, wide polyploids are unlikely to offer immediate advantages, and improvement is likely to derive from the controlled introgression of genes and gene complexes. There are exceptions, particularly in spreading the region of adaptation (e.g. triticale). At the other extreme in herbage plants, where intensive grassland utilization is relatively recent, management systems are now evolving to provide the new environments that novel allopolyploids might exploit.

From the genetic point of view we have to consider first gene expression, particularly dominance and epistatic relations, which determine the value of the hybrid, and second, gene transmission during reproduction, which affects fertility and stabilty. In sexual allopolyploids, the segregational properties of the genes depend on the 'pairing' relations of the homologous and homoeologous chromosomes. In outbreeders, useful hybrid combinations can be preserved even where the chromosome pairing is random through the effects of tetrasomic inheritance (Stebbins 1953; Breese & Thomas 1978), though this will not obtain under inbreeding (Lewis 1967). Further stability is achieved through preferential pairing of homologous chromosomes. This may arise even between closely related, relatively undifferentiated genomes, through genetic control mechanisms (Riley & Chapman 1958).

These general considerations only achieve real significance when related to a specific crop. Above all they indicate the need for an intimate knowledge of the crop and its requirements, and of the biology and genetic relations of the target species. Here allopolyploidy and its potential will be considered in relation to the herbage grasses with special reference to *Lolium* and *Festuca* spp.

Breeding for agronomic flexibility in grasses

The *Lolium*/*Festuca* complex of species is adapted to a wide range of conditions, and the four major agricultural species, *L. perenne*, *L. multiflorum*, *F. pratensis* and *F. arundinacea*, have complementary characteristics, which are recognized as useful agronomic attributes. At present the complementation of these attributes is possible only through the use of mechanical seeds mixtures. These mixtures are difficult to manage because of competitive interactions.

The four species are related to an extent that will allow hybridization, and thus genetic exchange, and it is a breeding aim to combine genetically their attributes through hybrids or their derivatives. We would thereby achieve sward flexibility through genotypic versatility rather than genetic heterogeneity.

Tetraploid *Lolium* hybrids

The two ryegrasses, *L. perenne* and *L. multiflorum*, which are diploid ($2n = 14$), are the most important grassland species in British agriculture. Together they offer a most useful complementation of characters and it has long been the aim to incorporate these characters in stable hybrids. The species cross easily to produce fertile diploid F_1s but in F_2 and later generations there is transgressive segregation leading to rapid genetic deterioration, including a degree of sterility (Naylor 1960).

Male-sterile clones have been used to produce commercial quantities of diploid F_1 seed (J. Joordens, personal communication). An allotetraploid programme was started as an alternative method, on the assumption that tetrasomic inheritance reinforced by a degree of preferential pairing would arrest genetic deterioration during seed multiplication. The programme was further advocated by the existence of valuable synthetic autotetraploid varieties

490   E. L. BREESE, E. J. LEWIS AND G. M. EVANS

of these species whose success derives from the larger cell size, bringing about reduced fibre and thus improved feed quality. We may thus hope to exploit favourable physiological (exophenotypic) effects of polyploidy together with stabilizing (endophenotypic) effects, giving wider adaptation through hybridity.

Genetic assumptions and breeding tactics

The consequences of tetrasomy and partial preferential pairing in terms of gene segregation have to be followed over enough generations (four or more) to achieve sufficient seed multiplication. First, consider a single gene locus with two alleles ($A:B$) in outbreeding populations,

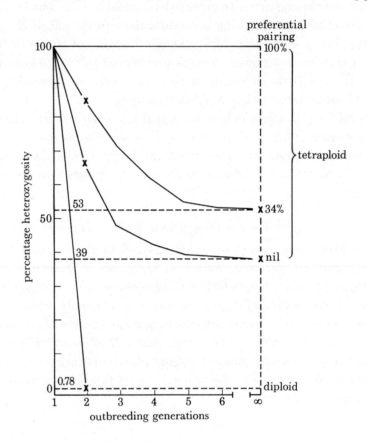

FIGURE 1. Gene segregation in diploid and tetraploid hybrids over successive outbreeding generations: percentage of population that are fully hybrid at seven loci, with two alleles ($A:B$), assuming no double reduction and that simplex, duplex and triplex (i.e. $AAAB$, $AABB$, $ABBB$) genotypes give the hybrid phenotype.

assuming random chromosome assortment (no double reduction) at the tetraploid level and that simplex, duplex and triplex (i.e. $AAAB$; $AABB$; $ABBB$) individuals confer the hybrid phenotype. In the diploid, equilibrium is achieved in the F_2 with 50% hybridity. In tetraploids, equilibrium is achieved exponentially after four or five generations. Then 87.5% of the population are hybrid with tetrasomic inheritance, increasing to 100% with increasing levels of preferential pairing (Breese & Thomas 1978). If we consider more loci, the disparity between the different inheritance patterns becomes more striking. Figure 1 gives the situation for seven unlinked loci. At the diploid level at equilibrium, only a small fraction of less than

1% are hybrid at all seven loci. In tetraploids, on the other hand, more than 40% of the population retain hybridity for all these loci, assuming tetrasomic inheritance, while this is increased to over 50% with the moderate degree of preferential pairing assumed. Breese (1981) has advanced comparisons to take into account combinations of these loci held in the hybrid state, and shows that up to 80% of the populations will retain hybridity for at least six loci by tetrasomic inheritance alone. Of course, amphiploidy will retain 100% hybridity at the duplex level.

If we equate the seven loci with the seven centromeres of the seven chromosomes of the haploid *L. multiflorum* and *L. perenne* genomes, we can see the potentially high capacity of the tetrasomic state alone to preserve hybrid combinations and how this can be considerably reinforced by preferential pairing. Of course, the situation is grossly oversimplified. For instance, we have to consider the gene linkages and recombination frequencies between the homoeologous chromosomes. All this can only be determined by experiment and experience.

The above considerations emphasize the need to select specific F_1 hybrid combinations for stabilization through tetrasomy. The breeding tactics were therefore the controlled paired crossing of individual genotypes (clones) from existing tetraploid varieties or newly induced tetraploid populations, followed by the selection of individual F_1s for fertility, stability and agronomic performance.

Fertility, uniformity and stability

In selected hybrids, relatively good fertilities were achieved despite fairly high levels of aneuploidy. In common with other 'raw' tetraploids, aneuploid gametes arise in the tetraploid hybrids from the irregular disjunction of quadrivalents and the presence of trivalents and univalents. In herbage crops a proportion can be tolerated since the less vigorous tend to be crowded out.

For individual plant characters the hybrids vary considerably in their uniformity, but usually they appear much more conservative of genetic variability than the diploid hybrids. The differences are less easy to quantify for individual characters in statistical terms because of the high errors attached to variances. Also the variances of the tetraploid families may be inflated by aneuploidy effects.

The stability of mean performance over generations also varies. For selected families, the change from F_2 to F_5 (equilibrium) is less than would be expected from tetrasomic inheritance and bespeaks a degree of preferential pairing as considered later.

Hybrid performance

Differences between F_1 families in performance as well as uniformity are marked (Breese *et al.* 1975). It will be appreciated that the method is largely empirical and has not so far involved critical experiment to determine the precise dominance and interactive properties at the tetraploid level of the genes controlling the different characters. Nevertheless, generalized comparisons of dominance relations can be made with respect to parental varieties of the different species. Thus in some of the hybrids the dominance of the Italian ryegrass parent is apparent up to the heading phase, when there is a switch to partial dominance of the perennial parent during recovery growth. This switch affects the canopy structure of the sward, which leads to more stable yields over frequent and infrequent defoliation than either parent and offers a possible greater flexibility for conservation and grazing managements.

Commercial varieties have now been produced that offer good conservation potential coupled with early and late grazing. They are highly palatable and have digestibility and intake characters that match or exceed the high quality of Italian ryegrass. Good winter hardiness and tolerance of drought and some diseases have been incorporated.

Future prospects

The hybrids thus show that it is possible to combine a plastic response to controllable (management) variables with low sensitivity to uncontrollable (environmental) variables. As pointed out, these first (primary) hybrids are single crosses between selected pairs of clones of the two parental species. Thus although hybridity is maximized between homoeologous chromosomes it is minimal between homologous chromosomes. Clearly the technique offers scope for the production of a wide range of such hybrids that are still better adapted to conservation and for grazing conditions. It allows rapid selection at the diploid level for complementary genotypes and uniting these in relatively stable tetraploids. But in fully exploiting this we need to know more of the dominance and epistatic relations of the genes controlling the various characters at both the diploid and tetraploid level so that we may reduce still further the empiricism currently involved in selecting parents and hybrids. But the tetraploid state offers the possibility of a second dimension of hybridity, that between homologues of the same genome, and thus affords further opportunities for enhancing heterosis and extending adaptability. This course, however, requires a fundamental reappraisal of the relations between chromosomes and the consequent segregational properties when more than two alleles are involved at a locus. For tetrasomy it can be shown that only 37% of a population at equilibrium is heterozygous for all four possible alleles at a single locus and only a very small proportion of the population is simultaneously fully hybrid for even three loci. It can also be shown that maximum hybridity cannot be advanced beyond 44% even for a single locus.

Apart from the inability to maximize hybridity with more than two alleles, tetrasomic segregation will also of course hamper the selection and stabilizing of desirable gene combinations in new varieties. Secondary hybridity will no doubt add to the genetic heterogeneity of the population, and by intercrossing our primary hybrid we can effectively create a new hybrid ryegrass species of the status of 4x *Dactylis glomerata* (cocksfoot). However, to continue to add more accurately to genotypic flexibility we require a more precise control of the chromosome pairing. We thus need to enquire more into the basis of stability and uniformity of the primary hybrids.

CHROMOSOME AFFINITIES IN *LOLIUM* HYBRIDS

Pairing preferences

The diploid hybrids between *L. multiflorum* and *L. perenne* regularly form bivalents and thus there is no indication of marked structural differences between the genomes. Nevertheless, the allotetraploid hybrids show on average an increased number of bivalents and a consequent reduction in multivalents compared with autotetraploids within the parental species. This suggests, but does not prove, a degree of preferential pairing.

More precise information in pairing preferences is now being obtained through the segregational patterns of enzyme variants (isozymes), which can be used as a means of genetically marking the different chromosomes. Using variants of the enzyme phosphoglucoisomerase (PGI/2), Breese & Thomas (1978) have estimated preferential pairing of the order of 34% among basic F_1 plants of the tetraploid hybrid cv. Sabrina. It would thus appear that the selection of this hybrid on the basis of its performance and stability had indeed involved a degree of preferential pairing, at least in the initial F_1 generation. To what extent this control is maintained in later generations is difficult to determine in this hybrid from the use of the PGI/2 locus. This is because we cannot completely identify parental chromosomes with isozyme allele for all basic plants. However, an analysis was carried out on samples from nine different seed lots of F_5/F_6 generations produced commercially in various parts of the country (E. W. Bean, unpublished). As expected, the data were too insensitive to detect differences between complete tetrasomic inheritance and partial preferential pairing, but they demonstrated that this level was significantly short of complete. The results were valuable in showing that the gene frequencies of the original F_1 plants had been consistently preserved over all seed lots, thus indicating that there had been no differential loss of parental genomes, for this marker at least, by selection or through aneuploidy. At the same time, through their genotypic frequencies, they demonstrated the need to increase preferential pairing before secondary hybridity could be usefully exploited.

Prospects for improving preferential pairing

Cytological evidence shows that the degree of bivalent formation varies markedly for different F_1s and would thus appear to be under genetic control. Most importantly, isozyme and other genetic markers have now been used to demonstrate that the degree of preferential pairing varies widely between different *L. multiflorum* × *L. perenne* hybrids for at least two loci (table 1).

More precise evidence of genes controlling variation in the degree of association of homoeologous chromosomes is available from hybrids between other species in the *Lolium* genus. Thus in an experimental cross between *L. temulentum* and *L. perenne* it was established not only that B chromosomes suppressed pairing in diploid hybrids and promoted bivalent pairing in the tetraploids, but that there were genes on the A chromosomes of *L. perenne* that had a similar and supplementary effect (Evans & Macefield 1972; Taylor & Evans 1977). Recent work by G. M. Evans & E. W. Davies (unpublished) with isoenzyme marker genes confirms that where bivalent association is enhanced in tetraploid hybrids it is between homologous chromosomes. These results are summarized briefly in table 2.

The usefulness of B chromosomes in breeding work is questionable owing to their non-mendelian inheritance and their dispensable nature. However, any variation on the A chromosomes is potentially useful and could possibly be utilized in breeding programmes.

There is thus evidence to suggest that useful genetic variation in terms of chromosome pairing control at the hybrid level exists within the parental species. A selection programme is now under way to use these genes in enhancing the degree of pairing in allotetraploids. This programme may also throw light on the nature, and hence the selective value, of such genes in the diploid parents since their most obvious effects are concerned with polyploid hybrids. Waines (1976) has advanced the argument that they may function as an isolating mechanism at the species level by reducing chromosome pairing and consequently the fertility level between diploid interspecific hybrids.

Incorporating pairing genes from other species

The stability and high fertility of triticale is dependent to a large degree on the action of the Ph gene from wheat. Not only does it suppress pairing between homoeologous wheat chromosomes but it also suppresses the synapsis of wheat and rye chromosomes. It is very likely that diploidizing mechanisms similar to the *Ph* locus in wheat occur in many other polyploid taxa. Although there are no known instances of natural polyploidy in the genus

TABLE 1. GENOTYPIC VARIATION IN LEVEL OF PREFERENTIAL CHROMOSOME PAIRING IN TETRAPLOID *L. PERENNE* × *L. MULTIFLORUM* CALCULATED FROM GENETIC SEGREGATION AT TWO LOCI

locus	range in percentage preferential pairing	
	random chromosome assort.	random chromatid assort.
anthocyanin 'c'	0–21.85	0–39.22
PGI/2	0–44.10	13.79–56.54

TABLE 2. CHROMOSOME PAIRING IN TETRAPLOID *L. TEMULENTUM* × *L. PERENNE* HYBRIDS

hybrids	bivalents per cell	preferential† pairing (%)
L. temulentum (Ba 6902) × *L. perenne* (clone A)		
0B	12.4	38.5
4B	13.0	84.4
L. temulentum (Ba 6902) × *L. perenne* (clone B)		
0B	11.5	0
4B	12.5	48.3

Note: (*a*) Ba 6902 is a homozygous line; (*b*) both *L. perenne* parents contained identical B chromosomes.

† Based on segregation at three unlinked isozyme loci.

Lolium itself, they abound in the closely related *Festuca*. Although a pairing control mechanism has not been clearly identified in these natural allopolyploids, there are strong indications of its presence (Adams & Allard 1977; Lewis *et al.* 1980; Jauhar 1975). As a long-term commitment, therefore, the identification and incorporation of such a mechanism into synthetic hybrids is indicated.

Either by selection within the species, or by incorporation of specific genes controlling homoeologous pairing, we thus have good prospects of producing a new amphidiploid ryegrass species with greater agronomic flexibility. It would then provide new opportunities for widening adaptability still further through secondary hybridity of homologous chromosomes. At this stage the incorporation of a third diploid genome from related *Festuca* spp. could be considered.

LOLIUM/FESTUCA HYBRIDS

Hybrids between Lolium spp. and F. pratensis

Meadow fescue (*F. pratensis*, $2n = 14$) is agriculturally far less important than the ryegrasses in the United Kingdom and complements these species in but a few characters. Nevertheless, the combination of these characters has been attempted by numerous workers (e.g. Jenkin 1933; Reusch 1959; Wit 1959; Hertzsch 1966), encouraged by the fact that natural hybrids that are very competitive are frequently found in certain habitats.

These *Festulolium* diploid hybrids, both natural and artificial, are highly sterile, which has been considered to be genic (Reusch 1960) or chromosomal (Ahloowalia 1965). Induced polyploidy restores fertility to a fairly high degree (Lewis 1972), and the formation of bivalents is higher than in the *Lolium* hybrids (table 3), suggesting a higher degree of preferential pairing. However, as shown in table 3, the level of bivalent association varies with the genotype, again indicating scope for improvement by selection (Osborne et al. 1977). A number of potential varieties of both amphiploids (i.e. *F. pratensis* hybridized with both *Lolium* species) show some complementation for late summer growth and good drought and winter hardiness (Lewis et al. 1973). Although markedly superior in yield to meadow fescue, these hybrids

TABLE 3. CHROMOSOME PAIRING IN AUTOTETRAPLOIDS AND ALLOTETRAPLOIDS IN *LOLIUM–FESTUCA*

	mean frequency	
	II	IV
L. perenne (4x)	6.5 (5.2–7.6)	3.4 (3.0–3.6)
L. perenne × L. multiflorum	9.0 (6.6–10.9)	2.1 (1.2–3.6)
L. perenne × F. pratensis	10.4 (8.8–11.8)	1.1 (0.6–1.6)
L. multiflorum × F. pratensis	10.8 (9.6–12.6)	0.8 (0.2–1.6)

have not so far matched the ryegrass parents or ryegrass hybrids under intensive U.K. conditions. Their tolerance of environmental stresses may, however, widen their adaptability into newer grassland areas. The amphiploids are also being backcrossed to the ryegrass parents with the aim of transferring the more desirable gene combinations of the fescue.

Hybrids between F. arundinacea *and* Lolium *spp.*

Next to the ryegrass hybrids, the combination of tall fescue with Italian ryegrass (*F. arundinacea* × *L. multiflorum*) would appear to offer the greatest complementation in characteristics, particularly for an increased, persistent conservation model. As such, it has received perhaps wider attention than most other interspecific hybrids (Terrell 1966). The hybrid with perennial ryegrass also has potentially high agronomic value.

F. arundinacea is an outbreeding hexaploid species ($2n = 42$) considered to be an amphidiploid, involving three basic diploid genomes (Lewis et al. 1980), with a close relationship to the *Lolium* spp. and *F. pratensis*. F_1 hybrids with the *Lolium* species are sterile, but fertility is restored in chromosome-doubled forms ($2n = 56$) (Buckner et al. 1961; Lewis 1966). The initial amphiploids have a high level of bivalent pairing, which is assumed to be preferential since the 28 chromosome undoubled hybrid shows a high number of multivalent associations and hence homoeologous pairings. It would thus appear that the genes controlling pairing in hexaploid tall fescue extend their influence to control pairing to some extent in the octoploid hybrid. However, the control is not absolute; a proportion of multivalents and univalents are formed, leading to chromosome loss and subsequent genetic instability (Lewis 1966). There is some parallel here with the instability of octoploid triticale considered earlier; this must raise the question of the hexaploid as an optimum number in these crop plants. In the Festuceae generally, however, octoploid and higher ploidy levels are not uncommon and this encourages the hope of developing more stable higher ploidy hybrids.

Although stably combining genomes of *Lolium* species with *F. arundinacea* has proved an elusive goal, polyploidy can be used as a genetic bridge for the transfer of genes and gene combinations between the parent species. Thus certain quality characteristics have been transferred from *L. multiflorum* into *F. arundinacea* (Buckner *et al.* 1977). In the United Kingdom we are attempting the transfer of valuable genes from the tall fescue to the ryegrass parent. The scope for this form of genetic transfer is enormous in the wide range of related *Lolium–Festuca* species.

In polyploid breeding, whether for combining genomes or genetic transfer, the limits are initially set by the numbers of hybrids that can be obtained. The crossability of the species varies markedly between different genotypes of donor species as well as between species. The development of tissue culture techniques for embryo culture and ultimately somatic hybridization (Cocking 1977) offers hopes of overcoming these obstacles (Dale 1976).

Conclusions

In this paper we have chiefly considered polyploidy as a means of genetically stabilizing interspecific hybrids in grasses. A number of features of the crop advocate such an allopolyploid breeding programme. First, there are unique opportunities for producing new and novel hybrids and their derivatives that are better adapted to intensive grassland systems than existing species. Secondly, since seed fertility and seed yield are not at a premium and a degree of non-uniformity may be tolerated, a range of polyploid techniques may be used. It is clear that polyploidy can be a valuable tool in combining genomes or facilitating genetic exchange, but its effective use depends on (i) defining realistic objectives in relation to the requirements and biology of the crop, (ii) a knowledge of the dominance/epistatic relations of the genes in the polyploid state, and (iii) an understanding of the nature and genetic control of homoeologous (intergenomic) chromosome pairing.

References (Breese *et al.*)

Adams, W. T. & Allard, R. W. 1977 Effect of polyploidy on phosphoglucose-isomerase diversity in *Festuca microstachys*. *Proc. natn. Acad. Sci. U.S.A.* **74**, 1652–1656.

Ahloowalia, B. S. 1965 Cytogenetic studies on natural hybrids between ryegrass and meadow fescue. *Z. Vererb-Lehre* **97**, 226.

Breese, E. L. 1981 Exploitation of genetic resources through breeding: *Lolium* species. In *Proc. Symp. on Genetic Resources of Forage Plants* (ed. R. A. Bray & J. G. McIvor). Melbourne: C.S.I.R.O. (In the press.)

Breese, E. L., Stephens, D. E. & Thomas, A. C. 1975 Tetraploid ryegrasses: allopolyploidy in action. In *Ploidy in fodder plants* (ed. B. Nuesch), pp. 46–49. Zurich: Eucarpia Report.

Breese, E. L. & Thomas, A. C. 1978 Uniformity and stability of *Lolium multiflorum* × *Lolium perenne* allotetraploids. In *Interspecific hybridization in plant breeding* (ed. E. Sanchez-Monge & F. Garcia-Olmedo) (Proc. 8th Eucarpia Congress, Madrid), pp. 155–160. Madrid: Ciudad Universitaria.

Buckner, R. C., Hill, H. D. & Burrus, P. B. 1961 Some characteristics of perennial and annual ryegrass × tall fescue hybrids and the amphiploid progenies of annual reygrass × tall fescue. *Crop Sci.* **1**, 75–80.

Buckner, R. C., Burrus, P. B. & Bush, L. P. 1977 Registration of 'Kenhy' tall fescue. *Crop. Sci.* **17**, 672–673.

Cocking, E. C. 1977 Plant protoplast fusion: progress and prospects for agriculture. In *Recombinant molecules: impact on science and society* (ed. R. F. Beers, Jr. & E. G. Bassett), pp. 195–208. New York: Raven Press.

Dale, P. J. 1976 Tissue culture in plant breeding. *Rep. Welsh Pl. Breed. Stn 1975*, pp. 101–115.

Dewey, D. R. 1980 Some application and misapplication of induced polyploidy to plant breeding. *In Polyploidy: biological relevance* (ed. W. H. Lewis), pp. 445–466. New York: Plenum.

Eigsti, O. J. & Dustin, A. P. 1955 *Colchicine in agriculture, medicine, biology and chemistry.* (470 pages.) Ames: Iowa State College Press.

Evans, G. M. & Macefield, A. J. 1972 Suppression of homoeologous pairing by B chromosomes in a *Lolium* species hybrid. *Nature, new Biol.* **236**, 110–111.

Hertzsch, W. 1966 Interspecific and intergeneric hybrids between *Lolium* and *Festuca*. In *Proc 10th Int. Grassl. Congr.*, Helsinki (ed. A. G. G. Hill), pp. 683–685. Helsinki: Finnish Grassland Association.

Jauhar, P. P. 1975 Genetic control of diploid-like meiosis in hexaploid tall fescue. *Nature, Lond.* **25**, 595–597.

Jenkin, T. J. 1933 Interspecific and intergeneric hybrids in herbage grasses. Initial crosses. *J. Genet.* **28**, 205–264.
Larter, E. N. 1976 Triticale. In *Evolution of crop plants* (ed. N. W. Simmonds), pp. 117–120. London: Longman.
Lewis, E. J. 1966 The production and manipulation of new breeding material in Lolium–Festuca. In *Proc. 10th Int. Grassl. Cong.*, Helsinki (ed. A. G. G. Hill), pp. 688–693. Helsinki: Finnish Grassland Association.
Lewis, E. J. 1972 Production of *Festuca/Lolium* hybrids. *Rep. Welsh Pl. Breed. Stn* 1971, p. 20.
Lewis, E. J., Humphreys, M. W. & Caton, M. P. 1980 Disomic inheritance in *Festuca arundinacea* Schreb. *Z. PflZücht.* **84**, 335–341.
Lewis, E. J., Tyler, B. F. & Chorlton, K. H. 1973 Development of *Lolium/Festuca* hybrids. *Rep. Welsh Pl. Breed. Stn 1972*, pp. 34–37.
Lewis, K. R. 1967 Polyploidy and plant improvement. *Nucleus* **10**, 99–110.
Naylor, B. 1960 Species differentiation in the genus *Lolium*. *Heredity, Lond.* **15**, 219–233.
Osborne, R. T., Thomas, H. & Lewis, E. J. 1977 *Lolium/Festuca* amphiploids. *Rep. Welsh Pl. Breed. Stn 1976*, p. 106.
Reusch, J. D. H. 1959 The nature of the genetic differentiation between *Lolium perenne* and *Festuca pratensis*. *S. Afr. J. Agric. Sci.* **2**, 271–283.
Reusch, J. D. H. 1960 The effects of gamma radiation on crosses between *Lolium perenne* and *Festuca pratensis*. *Heredity, Lond.* **14**, 51.
Riley, R. & Chapman, V. 1958 Genetic control of cytologically diploid behaviour of hexaploid wheat. *Nature, Lond.* **182**, 713–715.
Schwanitz, F. 1953 Die Zellgrösse als Grundelement in Phylogenese und Ontogenese. *Züchter* **23**, 17–44.
Stebbins, G. L. 1950 *Variation and evolution in plants.* New York: Columbia University Press.
Stebbins, G. L. 1953 Species hybrids in grasses. In *Proc. 6th Int. Grassl. Congr.* (ed. R. E. Wagner, W. M. Myers & S. H. Gaines), pp. 247–253. Washington: National Publishing Company.
Stebbins, G. L. 1956 Artificial polyploidy as a tool in plant breeding. In *Genetics in plant breeding* (Brookhaven Symposia in Biology, vol. 9), pp. 37–52. New York: Brookhaven National Laboratory.
Taylor, I. B. & Evans, G. M. 1977 The genetic control of homoeologous chromosome association in *Lolium temulentum* × *Lolium perenne* hybrids. *Chromosoma* **62**, 57–67.
Terrell, E. E. 1966 Taxonomic implications of genetics in ryegrasses (*Lolium*). *Bot. Rev.* **32**, 138–164.
Tigerstedt, P. M. A. 1978 Ecological aspects of adaptation in hybrid swarms of plants. In *Interspecific hybridization in plant breeding* (ed. E. Sanchez-Monge & F. Garcia-Olmedo) (*Proc. 8th Eucarpia Congr.*, Madrid), pp. 31–40. Madrid: Ciudad Universitaria.
Waines, J. G. 1976 A model for the origin of diploidizing mechanisms in polyploid species. *Am. Nat.* **110**, 415–430.
Wit, F. 1959 Hybrids of ryegrasses and meadow fescue and their value for grass breeding. *Euphytica* **8**, 1–12.

Haploidy and plant breeding

By J. G. T. Hermsen and M. S. Ramanna
Agricultural University, Department of Plant Breeding,
Lawickse Allee 166, Wageningen, The Netherlands

[Plate 1]

A haploid is an organism that looks like a sporophyte, but has the chromosome complement of a reduced gamete. There are several ways in which haploids can occur or be induced *in vivo*: spontaneously, mostly associated with polyembryony, and through abnormal processes after crosses, like pseudogamy, semigamy, preferential elimination of the chromosomes of one parental species, and androgenesis. In the crops described, haploids are or are near to being used in basic research and plant breeding.

The application of haploids in breeding self-pollinated crops is based on their potential for producing fully homozygous lines in one generation, which can be assessed directly in the field. Early generation testing of segregating populations is possible through haploids, because doubled haploids (DH) possess additive variance only. Haploids can also be applied in classical breeding programmes to make these more efficient through improved reliability of selection.

The application of haploids in cross-pollinated crops is also based on a rapid production of DH-lines, which can be used as inbred lines for the production of hybrid varieties. By means of haploids all natural barriers to repeated selfing are bypassed.

In autotetraploid crops there are two types of haploid. One cycle of haploidization leads to dihaploids; a second cycle produces monohaploids. The significance of dihaploids is in their greatly simplified genetics and breeding and in the possibility of estimation of the breeding value of tetraploid cultivars by assessing their dihaploids. The main drawback of dihaploids is their restriction to two alleles per locus. Also, after doubling, it is impossible to achieve tetra-allelism at many loci, the requirement for maximal performance of autotetraploid cultivars. Tetra-allelism can be obtained when improved dihaploids have a genetically controlled mechanism of forming highly heterozygous restitution gametes with the unreduced number of chromosomes. Monohaploids, after doubling or twice doubling, may lead to fully homozygous diploids and tetraploids. These are important for basic research, but not yet for practical application. Meiotic data of potato homozygotes at three ploidy levels are presented.

1. Origin and induction of haploids

A haploid is an organism that looks like a sporophyte, but has the chromosome complement of a reduced gamete. Haploids can originate *in vivo* or *in vitro*.

(a) In vivo *origin and induction*

Haploids may arise from the embryo sac in the following ways.

(i) Spontaneously, where it is often associated with polyembryony (for review see Lacadena 1974). High frequencies are found in oilseed rape, *Brassica napus* (Thompson 1969, 1974), and flax, *Linum usitatissimum* (Plessers 1963; Rajhathy 1976; Thompson 1977).

(ii) After a normal double fertilization with subsequent preferential elimination of the chromosomes of a specific genome in the early stages of embryo development (for reviews see Kasha 1974; Jensen 1975). Main examples are: barley, *Hordeum vulgare*, from interspecific hybridization with *H. bulbosum* (Lange 1969; Symko 1969; Kasha & Kao 1970); and wheat, *Triticum aestivum*, from intergeneric crosses with *Hordeum bulbosum* (Barclay 1975).

(iii) Through pseudogamy, which is the development of an unfertilized female gamete or cell after stimulation by the male nucleus (for reviews see Rowe 1974; Chase 1969). Important examples are: maize, *Zea mays*, from intervarietal crosses (Chase 1949); potato, *Solanum tuberosum*, from interspecific hybridization with *S. phureja* (Hougas et al. 1958); lucerne, *Medicago sativa*, from interspecific hybridization with *M. falcata* (Bingham 1969); tobacco, *Nicotiana tabacum*, from interspecific hybridization with *N. africana* (Burk et al. 1979); poplar, *Populus* species, from interspecific hybridization with mentor pollen (Stettler et al. 1969) or toluidine blue pollen treatment (Illies 1974); wheat cultivar Salmon with *Aegilops* cytoplasm after pollination with other cultivars (Tsunewaki et al. 1968).

(iv) Through semigamy, whereby reduced male and female gametes participate in embryogenesis but nuclear fusion does not occur; this process results in chimeral plants with sectors of maternal and paternal origin (for reviews see Turcotte & Feaster 1974; Choudhari 1978). An example is cotton, *Gossypium hirsutum* and *G. barbadense* (Turcotte & Feaster 1967), where a high haploid frequency was found in the doubled haploid line 57-4.

(v) Through androgenesis, whereby the maternal nucleus is eliminated or inactivated before fertilization of the egg cell and the haploid individual contains in its cells the chromosome set of the male gamete only (for review see Chase 1963). The frequency of this type is generally extremely low, but a dramatic increase in the frequency of androgenetic haploids occurred in the recessive maize mutant 'indeterminant gametophyte' (Kermicle 1969).

(b) In vitro *induction*

Haploids *in vitro* may be obtained by culturing anthers (for reviews see Sunderland 1974, 1980) and ovaries (for review see San Noeum 1978). A great deal of work has been done on anther culture since the early 1950s (Tulecke 1953). The first haploid plants were obtained in 1964 from *Datura innoxia* (Guha & Maheshwari 1964), closely followed by *Nicotiana tabacum* (Bourgin & Nitsch 1967) and the monocotyledon *Oryza sativa* (Niizeki & Oono 1968). By 1974, haploids from *in vitro* culture had been obtained in as many as 58 species (Sunderland 1974); this number is expanding rapidly. On the other hand, the number of species in which workable numbers can be produced at reasonable costs is still rather limited. The Solanaceae family compares favourably with most other plant families. Among the Gramineae the occurrence of undesirable albino plantlets from anther culture is common. As far as we are aware the haploid frequency is sufficiently high for practical application only in some species of *Nicotiana* and *Datura*. It is expected that this stage will, in due time, be reached in other genera.

Concerning *in vitro* haploid induction, Sunderland (1974, 1980) put forward a hypothesis of pollen dimorphism. Pollen *in vivo* contains a fraction of grains with a natural potential for embryogenesis. The proportion of this embryogenic pollen is under genetic and environmental control and is positively correlated with the yield of embryos *in vitro*. The incipient state of embryogenesis may also be induced in non-embryogenic pollen by pre-culture treatments, e.g. temperature stress, thus increasing the yield of embryos. Genetic control has been recognized by many researchers: large differences are found in embryogenic ability within a species and even within one segregating population. The effect of culturing on embryogenesis is influenced by many variables: the composition of the medium, the solid or liquid state of the medium, the stages of the microspores, the incubation conditions (light, temperature) and the nature of the material put into culture. Microspore stages are important in determining morphogenetic pathways. Two pathways can roughly be distinguished (Sunderland 1980), one leading to haploids via divisions of either the generative or the vegetative nucleus or of both, and the other leading to homozygous diploids and polyploids owing to fusion of pollen nuclei.

Both anther culture and haploid induction *in vivo* may give rise to heterozygous plants originating directly from so-called numerically unreduced ($=2n$) gametes. Different irregular meiotic events are known to lead to $2n$ gametes (see §2c).

2. Applications of haploids in plant breeding

The potential of haploids in plant breeding is dependent upon the nature of the crop: self-pollinated or cross-pollinated, diploid or autopolyploid.

(a) *Self-pollinated crops*

Self-pollinated crops are either diploids or allopolyploids. Pure allopolyploids are functional diploids and are often indicated as amphidiploids. They contain two genomes of each of two or three different species. The following crops are close to the use of haploids in breeding: the diploids barley, rice, *Datura* and flax, and the allopolyploids tobacco, cotton and wheat.

The basis of application is the so-called DH-line. A DH-line is a group of homozygous doubled haploids derived from a heterozygous diploid via monohaploids or amphimonohaploids. The potentials and advantages of using DH-lines in breeding may be summarized as follows.

(i) DH-lines enable the fastest possible production of homozygous lines, namely in one generation, whereas the conventional method of repeated selfings takes 6 or 7 generations.

(ii) DH-lines can be produced and reliably tested at any stage of a breeding programme from F_1, F_2 onwards to the purification of varieties.

(iii) DH-lines exhibit only additive and (additive × additive) variance, whereas early selfed generations contain a considerable proportion of non-selectable non-additive variance. Therefore DH-lines permit a more reliable selection in early generations and, in addition, an earlier and more reliable assessment of the breeding value of hybrid populations (Reinbergs *et al.* 1976). In view of possible undesirable linkages in the F_1, some breeders advocate the production of DH-lines in the F_2 or F_3.

(iv) DH-lines have the potential for improving the efficiency of existing breeding methods. So, in a recurrent selection method consisting of cycles of alternately intensive intercrossing and selfing of selected genotypes, selfing may well be replaced by DH-production because assessing DH-lines is much more reliable than assessing the selfed progeny of heterozygous plants.

(v) Whereas homozygous lines produced by repeated selfing are adapted to the climatic conditions of the region where they were produced, DH-lines have undergone no selection at all. This may be an advantage where breeding aims at producing varieties for regions with different climatic conditions (Kasha & Reinbergs 1975).

It may be concluded that the main advantages of the haploid method in breeding self-pollinated crops are a saving of time through fast homozygosis and an improvement of the efficiency of classical breeding methods through the improved reliability of selection.

(b) Cross-pollinated diploid crops

Just as in self-pollinators, the application of haploidy in cross-pollinated diploid crops is based on the use of DH-lines. However, owing to inbreeding depression, these lines cannot be used directly, but only as parental inbred lines for the production of hybrid varieties. The haploid method is being applied or is approaching the stage of application in the following crops: maize, asparagus, cacao, sugar beet, turnip, cole crops and poplar.

When inbred lines are being developed via haploids, all barriers to repeated selfing, which are characteristic of natural cross-pollinators, are bypassed, e.g. dioecy, self-incompatibility and long juvenile periods. An additional advantage for self-incompatible crops is that selection for weak incompatibility alleles or genetic backgrounds inhibiting the activities of such alleles is avoided.

The time saving is particularly apparent in biennial crops (asparagus, which is also dioecious) and in crops with a long juvenile period (cacao, poplar). Only via haploidy can inbred lines be developed in these crops. The latter holds true also for developing all-male hybrid varieties of asparagus, because for this purpose supermale (YY) lines are needed, which can only be obtained through *in vitro* culture of anthers from normal male (XY) plants. Male asparagus hybrids are generally superior to female ones.

When DH-lines are produced, a large variation in vigour is normally observed. In this connection two aspects should be considered. First, when large numbers of DH-lines can easily be produced, selection of lines with a relatively high vigour may be carried out, so that an economical production of large amounts of single-cross seeds is feasible. Secondly, weak inbred lines may have valuable genotypes. Such genotypes could be saved when weak haploids can be hybridized somatically *in vitro* and the selection of F_1 hybrids is possible in the test tube on the basis of hybrid vigour. Such F_1 hybrids may then be used either as parents in three-way or double crosses, or directly if they can be propagated vegetatively on a large scale.

(c) Autopolyploid crops $(2n = 4x)$

We do not think that a practical application of haploidy is possible in autohexaploid crops like timothy grass (and maybe sweet potato), because halving their chromosome number leads to sterile trihaploids. The situation is different in autotetraploid lucerne and common potato. In autotetraploid crops, two successive cycles of haploidization are basically possible and have in fact been realized in the potato (Van Breukelen *et al.* 1977); first from tetraploid $(2n = 4x = 48)$ to dihaploid $(2n = 2x = 24)$, and secondly from dihaploid to monohaploid $(2n = x = 12)$ (see figure 1). The first step is already a routine application; the second step is hampered by the generally very low frequencies of monohaploids as a consequence of (sub)-lethal genes, disharmonious genotypes in the 12-chromosome individuals, and the competition of the more vigorous plants from $2n$-gametes. For both steps the dominant marker embryo-spot in the pollinator *Solanum phureja* (Hermsen & Verdenius 1973) is being widely used and is indispensable for the second step.

Some aspects of autotetraploid crops influence the exploitations of haploids in breeding. Thus, autotetraploids are highly heterozygous, and deleterious recessive genes are frequent in many varieties. As already mentioned, this reduces the frequency of monohaploids. Also, breeding of autotetraploid crops is complicated by tetrasomic inheritance. This is particularly so since the highest possible level of heterozygosity (four different alleles per locus = tetra-allelism) is desirable for quantitative traits, especially for yield (Mendoza & Haynes 1974; Busbice & Wilsie 1966; Dunbier & Bingham 1975).

FIGURE 1. A representative leaf of potato cultivar Gineke (a) and of a dihaploid (b) and monohaploid (c) derived from that cultivar.

(i) *Dihaploids* $(2n = 2x)$

Obvious advantages of the application of dihaploids to breeding are the simplified genetics and breeding at the diploid level, the improved crossability with valuable diploid wild species which are indispensable, particularly in potato breeding, and finally the possibility of assessing dihaploids from cultivars (so-called 'gametic samples') to estimate the breeding value of each variety.

The great difficulties in working with dihaploids are a widespread occurrence of male sterility, a lowered female fertility and poor flowering of most dihaploids. In addition, self-incompatibility is introduced by haploidization, which hampers self-fertilization and may restrict crossability. A basic shortcoming of dihaploids is their restriction to two alleles per locus. For that reason we do not believe that diploid potatoes can ever compete with tetraploids.

If this is true, the breeder, after having obtained improved diploid genotypes, needs to have recourse to the tetraploid level in such a way that tetra-allelism is obtained at as many relevant loci as possible. Colchicine treatment or explant culture for doubling the chromosome number increase the coefficient of inbreeding and at best achieve di-allelism. A breeder therefore has to rely upon a genetically controlled meiotic abnormality leading to highly heterozygous restitution (or $2n$) gametes in the improved diploids. Such abnormalities result in the first instance from pseudohomotypic divisions (see figures 2 and 3, plate 1), in which there is no reduction division and no or hardly any genetic recombination, and secondly fused spindle formation at metaphase II of meiosis (see figures 4 and 5, plate 1), in which there is both reduction division and genetic recombination, the effect of which, however, largely disappears owing to nuclear restitution. These abnormalities, indicated as first division restitution (FDR), give rise to $2n$ gametes with a genotype nearly identical to that of the diploid parent.

The procedure in which these phenomena are used is called sexual or meiotic polyploidization. When a tetraploid cultivar is crossed with a diploid FDR genotype, unilateral sexual polyploidization occurs; if two diploid FDR genotypes are intermated, bilateral sexual polyploidization may take place. Both procedures may lead to highly heterozygous and vigorous tetraploid progeny.

The use of the FDR process in breeding is dependent on its inheritance. The fused spindle mechanism, which in some clones occurs in 70–80% of the pollen mother cells (see figure 4), is controlled by one recessive gene, according to Mok & Peloquin (1975). Iwanaga (1980) found the average frequency of that gene in 51 potato cultivars to be 70%. This high frequency suggests either a selective advantage of that gene or a role of that gene in the evolution of the potato. Similar genes may have played a role in the evolution of many other genera in which the restitution mechanism has been detected (Harlan & De Wet 1975).

(ii) *Monohaploids* $(2n = x)$

Monohaploids from an autotetraploid crop are the only tool for the production of homozygous diploid and tetraploid genotypes of such a crop. Briefly, the procedure is twice haploidizing followed by twice doubling.

$$\underbrace{\text{cultivar} \to \text{dihaploid}}_{\text{heterozygous}} \to \underbrace{\text{monohaploid} \to \text{DH} \to \text{doubled DH}}_{\text{homozygous}}$$

In potato monohaploids, DH clones and doubled DH clones have been produced, and some could be studied cytologically. More than 90% of the 2247 pollen mother cells of 31 monohaploids studied had univalents only. In the 181 remaining pollen mother cells, the pairing configurations were 1, 2, 3 or even more bivalents, and in one cell a trivalent occurred. Clear differences in the amount of pairing were found between the monohaploids studied.

Description of plate 1

FIGURES 2 AND 3. First division restitution through pseudohomotypic division in the potato clone Y 17. Bar is 10 μm

FIGURE 2. Upper cell mainly with univalents, lower cell with most univalents dividing mitotically.

FIGURE 3. A group of cells in different meiotic stages: metaphase I, anaphase and telophase I. Typical second division stages are lacking in this clone and dyads are formed at the end.

FIGURES 4 AND 5. First division restitution through fused spindles in the potato clone IV P 10. Bar is 10 μm.

FIGURE 4. Metaphase II of meiosis showing cells with fused spindles (the majority) and cells with non-fused spindles that are either parallel (arrows) or at an angle (asterisk).

FIGURE 5. Dyads formed as a result of spindle fusion.

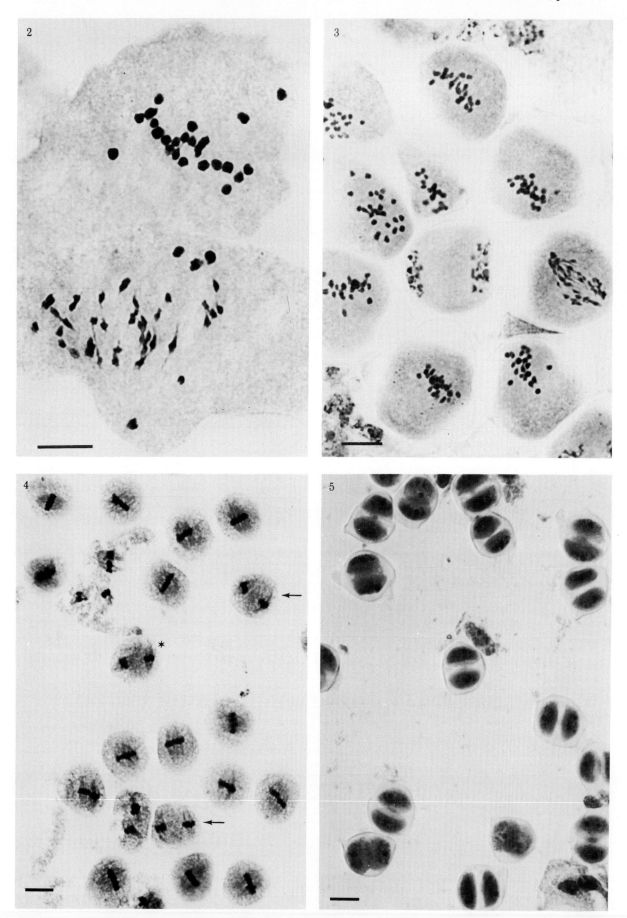

FIGURES 2–5. For description see opposite.

Meiosis was also studied in eight double monohaploids. In six of them meiosis was nearly normal (in most cells 12 bivalents), but in two a high frequency of univalents was found, which may have been caused by desynapsis.

DH clones were doubled to the tetraploid level, and four such homozygous tetraploids were studied meiotically. The number of quadrivalents in three of the tetraploidized monohaploids was very high (6.1–10.5 per cell) compared with that in cultivar Gineke (2.4 per cell). One tetraploid clone behaved similarly to its diploid parent in showing desynapsis. The high average number of quadrivalents in most homozygous autotetraploids compared with cultivars might indicate that potato cultivars are not fully autotetraploid and display a decreased homology between their genomes.

The potential of monohaploids for potato breeding is very limited. Unless monohaploids can be produced in large numbers and somatic hybridization of monohaploids becomes routine, the use of monohaploids for breeding hybrid varieties does not make sense.

The new idea of growing potatoes from true seeds, which despite disadvantages has a number of striking advantages, especially for use in developing countries, has been associated with the production of homozygous genotypes and F_1 hybrids to improve the uniformity of the crop. However, with the knowledge and material available, the idea may at present only be realized by exploiting the FDR mechanisms mentioned previously.

3. Conclusion

The lack of methods for an efficient and cheap production of large numbers of haploids has limited the number of crops in which haploids can be used for breeding purposes. There are worldwide interests and activities in the field of induction and use of haploids in basic research and breeding. It may be expected that in the decades ahead rapid progress will be made.

References (Hermsen & Ramanna)

Barclay, I. R. 1975 High frequencies of haploid production in wheat (*Triticum aestivum*) by chromosome elimination. *Nature, Lond.* **265**, 410–411.

Bingham, E. T. 1969 Haploids from cultivated alfalfa, *Medicago sativa* L. *Nature, Lond.* **221**, 865–866.

Bourgin, J. P. & Nitsch, J. P. 1967 Obtention de *Nicotiana* haploides à partir d'étamines cultivées in vitro. *Annls Physiol. Vég.* **9**, 377–382.

Breukelen, E. W. M. van, Ramanna, M. S. & Hermsen, J. G. T. 1977 Parthenogenetic monoploids ($2n = x = 12$) from *Solanum tuberosum* L. and *S. verrucosum* Schlechtd. and the production of homozygous potato diploids. *Euphytica* **26**, 263–271.

Burk, L. G., Gerstel, D. U. & Wernsman, E. A. 1979 Maternal haploids of *Nicotiana tabacum* L. from seed. *Science, N.Y.* **206**, 585.

Busbice, T. H. & Wilsie, C. P. 1966 Inbreeding depression and heterosis in autotetraploids with application to *Medicago sativa* L. *Euphytica* **15**, 52–67.

Chase, S. S. 1949 Monoploid frequencies in a commercial doublecross hybrid maize, and in its component singlecross hybrids and inbred lines. *Genetics, Princeton* **34**, 328–332.

Chase, S. S. 1963 Androgenesis – its use for transfer of maize cytoplasm. *J. Hered.* **54**, 152–158.

Chase, S. S. 1969 Monoploids and monoploid-derivatives of maize (*Zea mays* L.). *Bot. Rev.* **35**, 117–167.

Choudhari, H. K. 1978 The use of semigamy in the production of cotton haploids. *Bull. Torrey bot. Club* **105**, 98–103.

Dunbier, M. W. & Bingham, E. T. 1975 Maximum heterozygosity in alfalfa: results using haploid-derived autotetraploids. *Crop. Sci.* **15**, 527–531.

Guha, S. & Maheshwari, S. C. 1964 In vitro production of embryos from anthers of *Datura*. *Nature, Lond.* **204**, 497.

Harlan, J. R. & De Wet, J. M. J. 1975 On O. Winge and a prayer: the origins of polyploidy. *Bot. Rev.* **41**, 361–390.

Hermsen, J. G. T. & Verdenius, J. 1973 Selection from *Solanum tuberosum* Group *Phureja* of genotypes combining high-frequency haploid induction with homozygosity for embryo-spot. *Euphytica* **22**, 244–259.

Hougas, R. W., Peloquin, S. J. & Ross, R. W. 1958 Haploids of the common potato. *J. Hered.* **47**, 103–107.

Illies, Z. M. 1974 Induction of haploid parthenogenesis in aspen by post-pollination treatment with toluidine blue. *Silvae Genet.* **23**, 221–226.

Iwanaga, M. 1980 Diplogynoid formation in diploid potatoes. Ph.D. thesis, University of Wisconsin.

Jensen, C. J. 1975 Barley monoploids and doubled monoploids: techniques and experience. In *Barley genetics* (ed. H. Gaul), vol. 3, pp. 316–345. München: Verlag Karl Thiemig.

Kasha, K. J. & Kao, K. N. 1970 High frequency haploid production in barley (*Hordeum vulgare* L.). *Nature, Lond.* **225**, 874–876.

Kasha, K. J. 1974 Haploids from somatic cells. In *Haploids in higher plants* (ed. K. J. Kasha), pp. 67–87. University of Guelph.

Kasha, K. J. & Reinbergs, E. 1975 Utilization of haploids in barley. In *Barley genetics* (ed. H. Gaul), vol. 3, pp. 307–315. München: Verlag Karl Thiemig.

Kermicle, J. L. 1969 Androgenesis conditioned by a mutation in maize. *Science, N.Y.* **166**, 1422–1424.

Lacadena, J. R. 1974 Spontaneous and induced parthenogenesis and androgenesis. In *Haploids in higher plants* (ed. K. J. Kasha), pp. 13–32. University of Guelph.

Lange, W. 1969 Cytological and embryological research on crosses between *Hordeum vulgare* and *H. bulbosum*. *Versl. Landb. Onderz. Ned.* no. 719. (162 pages). (In Dutch.)

Mendoza, H. A. & Haynes, F. L. 1974 Genetic basis of heterosis for yield in the autotetraploid potato. *Theor. appl. Genet.* **45**, 21–25.

Mok, D. W. S. & Peloquin, S. J. 1975 The inheritance of three mechanisms of diplandroid (2n-pollen) formation in diploid potatoes. *Heredity, Lond.* **35**, 295–302.

Niizeki, U. & Oono, K. 1968 Induction of haploid rice plants from anther culture. *Proc. Japan Acad.* **44**, 554–557.

Plessers, A. G. 1963 Haploids as a tool in flax breeding. *Cereal News* **8**, 3–6.

Rajhathy, T. 1976 Haploid flax revisited. *Z. PflZücht.* **76**, 1–10.

Reinbergs, E., Park, S. J. & Song, L. S. P. 1976 Early identification of superior barley crosses by the doubled haploid technique. *Z. PflZücht.* **76**, 215–224.

Rowe, P. R. 1974 Methods of producing haploids: parthenogenesis following interspecific hybridization. In *Haploids in higher plants* (ed. K. J. Kasha), pp. 43–52. University of Guelph.

San Noeum, L. H. 1978 In vitro induction of gynogenesis in higher plants. In *Proc. Conf. Broad Genet. Base Crops* (ed. A. C. Zeven & A. M. van Harten), pp. 327–330. Wageningen: Pudoc.

Stettler, R., Bawa, K. & Livingston, G. 1969 Experimental induction of haploid parthenogenesis in forest trees. In *Induced mutations in plants*, pp. 611–619. Vienna: I.A.E.A.

Sunderland, N. 1974 Anther culture as a means of haploid induction. In *Haploids in higher plants* (ed. K. J. Kasha), pp. 91–122. University of Guelph.

Sunderland, N. 1980 Anther and pollen culture 1974–1979. In *The plant genome* (ed. D. R. Davies & D. A. Hopwood), pp. 171–183. Norwich: The John Innes Charity.

Symko, S. 1969 Haploid barley from crosses of *Hordeum bulbosum* (2x) × *H. vulgare* (2x). *Can. J. Genet. Cytol.* **11**, 602–608.

Thompson, K. F. 1969 Frequencies of haploids in spring oil-seed rape (*Brassica napus*). *Heredity, Lond.* **24**, 318–319.

Thompson, K. F. 1974 Homozygous diploid lines from naturally occurring haploids. In *Proc. 4 Internat. Rapskongr.*, pp. 119–124.

Thompson, K. F. 1977 Haploid breeding technique for flax. *Crop. Sci.* **17**, 757–760.

Tsunewaki, K., Noda, K. & Fujisawa, T. 1968 Haploid and twin formation in a wheat strain Salmon with alien cytoplasm. *Cytologia* **33**, 526–538.

Tulecke, W. R. 1953 A tissue derived from the pollen of *Ginkgo biloba*. *Science, N.Y.* **117**, 599–600.

Turcotte, E. L. & Feaster, C. V. 1967 Semigamy in Pima cotton. *J. Hered.* **58**, 54–57.

Turcotte, E. L. & Feaster, C. V. 1974 Semigametic production of cotton haploids. In *Haploids in higher plants* (ed. K. J. Kasha), pp. 53–64. University of Guelph.

Discussion

J. HESLOP-HARRISON, F.R.S. (*Welsh Plant Breeding Station, Aberystwyth, U.K.*). Professor Hermsen has described an interesting mechanism for the formation of diploid gametes. As I understand his account, diploidization occurs by the fusion of metaphase plates after meiosis I, an event observed as an occasional aberration in many dicotyledons. However, his micrographs show an unusually high proportion of meiocytes with parallel meiosis II spindles. In dicotyledons in general, the spindles of the second division tend always to be at right angles – this is a prerequisite, indeed, for the cleavage that gives the tetrahedral disposition of the spores in the tetrad. Would one be right in supposing that the genetic stocks that he is dealing with have an abnormally high incidence of second-division cleavage anomalies?

J. G. T. HERMSEN. In this particular genotype high frequencies of fused spindles and $2n$ pollen are found. There are other plants, however, with far fewer $2n$ pollen grains. This indicates that the expressivity of the responsible gene varies according to the genotype. This is quite common with meiotic mutants. The gene appears to be present in many genotypes of many species.

A. P. M. DEN NIJS (*Institute for Horticultural Plant Breeding, Wageningen, The Netherlands*). I should like to underline Professor Hermsen's statement that these gametes are of frequent occurrence. Searches for $2n$ pollen and $2n$ eggs in many diploid species of the *Solanum* polyploid complex by several workers have revealed substantial numbers of plants producing a moderate to high percentage of $2n$ pollen or $2n$ eggs in addition to normal gametes. One single recessive character, *parallel spindles* (*ps*), has been found to be responsible for the production of FDR $2n$ pollen. Genetic control of $2n$ egg production has been indicated. The *ps* allele frequently occurs in many accessions of various diploid species, suggesting a role for this gene in sexual polyploidization during the evolution of the polyploid complex of the tuber-bearing Solanaceae. If so, the gene should also be encountered in polyploid species and it has indeed often been found in *Solanum tuberosum* ($2n = 4x$), for example, and in dihaploids derived from it. The occurrence of $2n$ gametes facilitates gene flow from the diploid level to the tetraploid and even the hexaploid level by unilateral (e.g. $4x \times 2x \to 6x$) and bilateral (e.g. $2x \times 2x \to 4x$) sexual polyploidization.

J. G. T. HERMSEN. The role of $2n$ gametes in the evolution of polyploid complexes has been suggested by several authors and reviewed by Harlan & De Wet (1975). Regarding the genus *Solanum*, Dr den Nijs referred to the recessive gene controlling parallel spindles (*ps*). Dr Ramanna and I believe that only fused spindles invariably lead to FDR gametes, whereas parallel spindles are incidental orientations of the two metaphase II spindles, which may or may not lead to FDR gametes (see figure 4). Professor Heslop-Harrison made the interesting suggestion that the *ps* gene might well cause a random orientation of second metaphase spindles and this might lead to a greater chance of fused spindles compared with the normal orientation at an angle of 60° in *Solanum* genotypes not having *ps*. However, this hypothesis can hardly explain the high frequency of pollen mother cells with fused spindles.

To return to the original question of Professor Heslop-Harrison about cleavage anomalies, Dr M. S. Ramanna has found that the parallel orientation of metaphase II spindles, although theoretically unexpected, may occur at very high frequencies in normal tetrad-forming genotypes. Ramanna's interpretation is that, regardless of their orientation at metaphase II, the spindle axes are reoriented through the formation of the secondary phragmoplasts at the end of telophase II. This phenomenon is well known in dicotyledonous angiosperms with simultaneous cytokinesis during the second division of meiosis. We have diploid potato clones that produce nearly 100 % dyads instead of tetrads. If we suppose that dyad formation is a consequence of abnormal cleavage, these clones must have a high incidence of cleavage abnormalities.

Intraspecific chromosome manipulation

By C. N. Law, J. W. Snape and A. J. Worland

Plant Breeding Institute, Maris Lane, Trumpington, Cambridge CB2 2LQ, U.K.

Whole chromosome manipulation within a crop species is restricted in the main to the hexaploid bread wheat, *Triticum aestivum* ($2n = 6x = 42$). Directed manipulation within this group is possible because aneuploidy is tolerated, and at least 60 monosomic sets, representative of many of the successful wheats of the world, are available. These permit, after recurrent backcrossing and cytological selection, the substitution of whole chromosomes from one variety into another.

Substitution lines have been used to identify chromosomes and genes responsible for varietal differences affecting a range of important agronomic characters, and provide the most efficient method of genetical analysis available in wheat or indeed in any crop species. The length of time taken to develop substitution lines is, however, a weakness of the approach. A method in which monosomic sets are crossed reciprocally overcomes this weakness and permits the identification of significant chromosomal effects in one to two generations. Modifications to the method enable chromosomal effects to be studied in any varietal combination even though monosomic sets are not available in all the varieties being studied.

The results obtained by intraspecific chromosome manipulation have been exploited in the development of a high-yielding spring wheat after recombination of a single chromosome substituted into a winter wheat. This example suggests that the methods may have direct applications to plant breeding where large chromosomal effects have been identified. This is particularly relevant where the variation concerned is difficult to assess.

These methods also have a major part to play in increasing the understanding of the genetic architecture of wheat. It is also likely that they will have value in transferring useful genes from distantly related varieties.

Introduction

In the main, whole chromosome manipulation within a crop species is confined to the polyploids, where the duplication of genes permits the loss of chromosomes without disastrous consequences on a plant's phenotype. Such aneuploids have been assembled in a number of polyploid crop species, e.g. cotton, oats and wheat (Brown 1966; Rajhathy & Thomas 1974). However, it is in the latter species that the use of aneuploids has been the most extensive, and an elaborate methodology has evolved for their use in genetical analysis and chromosome manipulation.

The bread wheat of agriculture, *Triticum aestivum* ($2n = 6x = 42$), is a hexaploid species, and a range of aneuploids (monosomics, tetrasomics, ditelocentrics, nullisomics) representing each of the 21 chromosomes of wheat have been assembled over the years. This collection was first made in the wheat variety Chinese Spring by Sears (1954) in the United States, but more recently complete sets of aneuploid lines have been systematically obtained in a number of varieties (Law & Worland 1973).

Inter-varietal chromosome substitutions

The ability to tolerate the loss of a chromosome and to produce cytologically recognizable marker chromosomes (telocentrics) has enabled the substitution of single chromosomes from a donor variety for their homologues in a recipient variety (Sears 1953; Unrau 1950). The method is illustrated in table 1.

By recurrent backcrossing to the monotelocentric line of the recipient variety and selecting in each generation for monosomic plants having only complete chromosomes, it is possible to ensure that the donor chromosome is maintained intact. After a number of backcrosses, the

TABLE 1. DEVELOPMENT OF AN INTER-VARIETAL CHROMOSOME SUBSTITUTION LINE OF CHROMOSOME 1A BY USING A MONO-TELOCENTRIC LINE IN VARIETY A AS THE RECURRENT PARENT

(Selection is practised for monosomic plants after each hybridization. In this way the substituted chromosome is maintained intact and in the hemizygous condition until disomics are selected in the final selfing generation.)

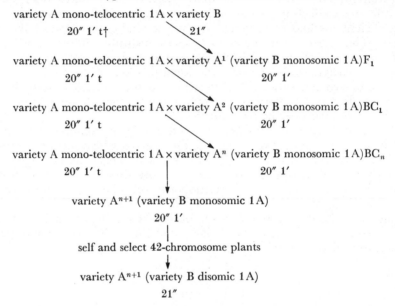

† $''$ refers to a bivalent and $'$ refers to a univalent at first metaphase of meiosis.

background will either be genetically identical or close to the recipient variety, so that by selfing, a disomic substitution line can be obtained that will be true-breeding and which can be treated as a normal variety in assessing the effect of the substituted chromosome. Since appropriate aneuploids exist for each of the 21 chromosomes of wheat, 21 inter-varietal chromosome substitutions are possible for each pair of recipient and donor varieties.

Inter-varietal chromosome substitutions provide one of the best means of studying quantitative characters. The power of the method may be illustrated by referring to a recent study of the genetics of the semi-dwarfism that occurs in some Mediterranean wheats, for example the varieties Mara and Sava.

Semi-dwarfism has, of course, been a major objective of wheat breeders for many decades, and the introduction of the dwarfing genes $Rht1$ and $Rht2$ from the Japanese variety Norin 10 to produce the 'green revolution' wheats has been one of the major achievements of plant breeding (Borlaug 1968). However, the dwarfism of Norin 10 is genetically different from the dwarfism of Mediterranean wheats. There is therefore a need to understand the nature of this difference more clearly.

In crosses between the tall, W. European variety, Cappelle-Desprez and the Italian semi-dwarf variety, Mara, the F_2 segregation for height is continuous, suggesting that genes of large effect are not involved in this varietal difference. A series of chromosome substitution lines in which Mara chromosomes replace their homologues in Cappelle-Desprez have been produced, and the variation in height between these substitutions lines is shown in figure 1. This clearly indicates that chromosome 2D and a gene or genes on the short arm of 5B and/or 7B ($5B^s$–$7B^s$) are responsible for the semi-dwarfism of Mara.

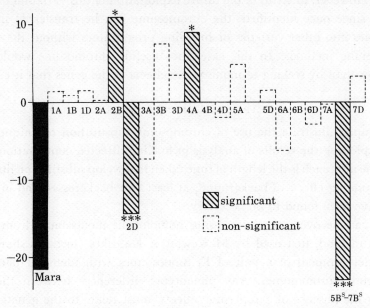

FIGURE 1. Deviations in final plant height from Cappelle-Desprez (103 cm) of chromosome substitution lines of Mara into Cappelle-Desprez. Because Cappelle-Desprez carries the reciprocal translocation, $5B^L$–$7B^L$ and $5B^s$–$7B^s$, not present in Mara, it has not been possible to substitute the complete 5B or 7B chromosome into Cappelle-Desprez. The $5B^s$–$7B^s$ substitution line therefore refers to the transfer of only the short arms of 5B and 7B of Mara. The long arms of these two chromosomes could not be transferred because the $5B^L$–$7B^L$ monosomic of Cappelle-Desprez has yet to be obtained. *, $0.05 > p > 0.01$; ***, $p < 0.001$.

This analysis of course refers to whole chromosomes, so that it is possible that more than one gene is responsible for these major chromosomal effects and that these genes are loosely linked. The breakdown of these linkages might explain the absence of discontinuities in the F_2 generation. Such a possibility may be tested by crossing each of the critical substitution lines to Cappelle-Desprez and observing the F_2 products of this single chromosome segregation on a non-segregating background. This has not been done yet, although the method has been used successfully in other cases (Law 1966; Law et al. 1978). However, whatever the outcome, the results of this substitution exercise clearly demonstrates the power of inter-varietal chromosome substitutions in genetical analysis.

Height is, of course, an easy character to measure, but this is not so for many of the characters of interest to the wheat breeder. More often than not genetical estimates must be based upon large numbers of replicates and the use of elaborate techniques of estimation that are often time consuming and tedious. Under these circumstances, inter-varietal chromosome substitutions are again more efficient than conventional methods of genetical analysis.

One example of this advantage is the study of the genetical control of the eye-spot resistance of Cappelle-Desprez. This fungal disease is assessed by artificially infecting juvenile wheat plants with the fungus *Cercosporella herpotrichoides* and scoring the degree of penetrance of the fungus through the leaf sheaths. To detect resistance to the fungus, extensive replication is required, so that the method is labour-intensive and does not readily lend itself to easy selection or to detailed genetical studies by conventional procedures.

On the other hand, a study of the substitution lines in which Cappelle-Desprez chromosomes replace their homologues in the susceptible variety Chinese Spring indicated that the resistance of Cappelle-Desprez was controlled mainly by one chromosome, chromosome 7A (Law *et al.* 1976). Again, it is possible that this effect could be due to a number of genes concentrated on this chromosome. However, in terms of the future exploitation of this variation in breeding this need not matter, since once identified, the chromosome can be transferred intact by cytogenetical techniques into other varieties or breeding programmes without the need to resort to laborious screening methods. In this case, the 'useful' chromosome would be followed cytologically rather than by seeking to monitor the effects of the genes that it carries.

Reciprocal monosomics

These two examples illustrate the use of chromosome substitution techniques in genetical analysis and in exploiting the results of analysis in making directed contributions to breeding. The weakness of the approach is the length of time taken to develop substitution lines. To remove the possible confounding effects of background, at least four backcrosses, and in some cases as many as eight, have been found to be necessary.

This weakness can be overcome by crossing homologous monosomics from two varieties reciprocally. This method, first used by McKewan & Kaltsikes (1970), is shown in figure 2 and depicts the development of a pair of F_1 monosomics with identical backgrounds but different hemizygous chromosomes. Any phenotypic differences between these reciprocal hybrids, assuming the absence of cytoplasmic effects, must relate to the genetical differences between the two hemizygous chromosomes. Because large numbers of F_1 monosomic hybrids may not be easy to produce, assessment may be deferred to the F_2 generation, where the means of the two reciprocally derived F_2 populations may be compared. Disomics may also be selected in the F_2 generation so that those genes ineffective in the hemizygous state may be detected and their effects estimated.

The reciprocal monosomic method will permit the detection of chromosomes whose effects are large in relation to the segregation of genes on other chromosomes. The method will therefore identify those chromosomes that are likely candidates for the future development of substitution lines and for directed introduction into breeding programmes.

Two examples of this approach will suffice. The first concerns the different chromosomal control of ear-emergence time between the spring/winter variety Bersée and the winter wheat Cappelle-Desprez. The results of crossing the monosomics of these varieties reciprocally and scoring the F_2 monosomic populations in the field after a spring sowing are shown in figure 3.

Clearly a number of chromosomes are responsible for the differences between the two varieties, so that the character can be regarded as polygenic. However, a major proportion of the variation is associated with chromosomes of homoeologous group 5 and 7, the major effect being determined by chromosome 5A. It is known from other aneuploid studies that 5A carries *Vrn*1, a gene for vernalization requirement, which has a major effect on the control of winter and spring wheat differences (Halloran & Boydell 1967; Law *et al.* 1976; Snape *et al.* 1976).

The second example concerns the chromosomal location of *Rht*2, one of the dwarfing genes from Norin 10. Although the presence of this gene can now be detected relatively easily in plants by their lack of response to exogenous gibberellic acid (Gale & Marshall 1973), it nevertheless provides an excellent example of the effectiveness of the reciprocal monosomic method in detecting major effects on a quantitative character. Disomic F_2 plants were selected from each of the reciprocal crosses between monosomics 4D of Bersée and the semi-dwarf wheat Hobbit, and their final plant heights obtained. These indicated that those plants carrying 4D of Hobbit were on average 20% shorter than plants carrying the Bersée homologue, thus confirming the location of Rht2 on Hobbit 4D (Gale et al. 1975).

However, once a chromosomal effect of this magnitude has been detected among reciprocally derived F_2 disomics, it is then but a simple matter to look at F_3 lines to study the consequence of such a chromosomal effect on other characters. A number of F_3 lines from the Bersée 4D with

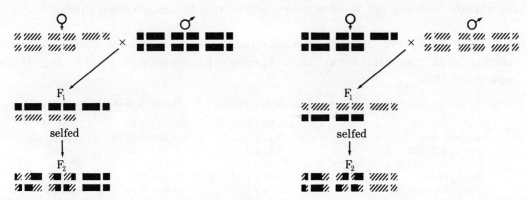

FIGURE 2. The development of F_1 monosomic hybrids with identical backgrounds but different hemizygous chromosomes, by crossing, reciprocally, homologous monosomics from two different varieties. For the sake of simplification, only three pairs of chromosomes are depicted instead of the normal 21. Different varietal chromosomes are shown as either hatched or in black.

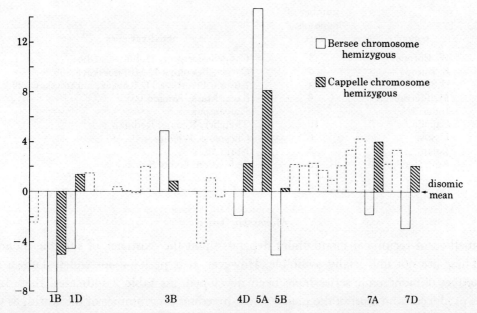

FIGURE 3. Mean ear-emergence times of F_2 reciprocal monosomic hybrids between Bersée and Cappelle-Desprez defined from the controlled disomic F_2 mean. Histograms of reciprocal monosomic pairs that are significantly different from each other are drawn with solid lines, and the chromosomes responsible for these differences are given at the bottom of the figure. (Law et al. (1979).)

Hobbit 4D cross were therefore grown; the different effects of the 4D chromosomes on a number of important agronomic characters are shown in table 2. Apart from the large effect on height, the effects on two of the yield components, grain number and grain size, were noteworthy. However, the effects were opposed so that their combined effect on overall yield was not significant.

Again, in these types of experiment it is not possible to state whether the effects are due to one or more genes, or that the same genes are affecting each of the characters. In the present case, after more detailed studies it was found that the effects are all due to the gene for dwarfism, $Rht2$ (Gale 1978). Obviously for other situations, the numbers of genes and their relations between each other would need to be studied. However, in many cases the genetic associations will be such that the breakages of linked complexes will not be necessary, and the major benefits in plant breeding will be obtained by transferring the chromosome intact.

TABLE 2. MEANS OF F_3 LINES, HOMOZYGOUS FOR CHROMOSOME 4D OF EITHER BERSÉE OR HOBBIT, DERIVED FROM THE RECIPROCAL CROSS BETWEEN BERSÉE MONOSOMIC 4D AND HOBBIT MONOSOMIC 4D

	Bersée 4D	Hobbit 4D	difference
height/cm	106.80 ± 2.07	88.45 ± 2.47	+18.35***
tiller number	8.41 ± 0.21	8.77 ± 0.21	−0.36
spike length/cm	9.92 ± 0.17	9.99 ± 0.20	−0.07
percentage fertility 1 × 2 florets	79.97 ± 1.09	82.64 ± 1.19	−2.67
grains per ear	60.89 ± 1.42	65.79 ± 2.20	−4.90
1000 grain mass/kg	55.14 ± 1.00	48.87 ± 0.99	+6.27***
grain mass per plant/g	23.20 ± 0.60	21.47 ± 1.31	+1.73

***, $p < 0.001$.

TABLE 3. MONOSOMIC SERIES IN WHEAT

region	number of monosomic sets	noted varieties
W. Europe	11	Cappelle-Desprez, Hobbit, Caribo
E. Europe	7	Carola, Besostaya I, Mironovskaya 809
U.S.S.R.	14	Aurora, Besostaya II, Kavkaz, Saratovskaya 29
Mediterranean	6	Sava, Mara, Aragon 03
India	2	Kalyansona
Canada	10	Thatcher, Rescue, Redman
U.S.A.	6	Cheyenne, Wichita
Japan	3	Norin 10
Australia	3	Federation, Gabo, Spica
total	62	

Monosomic series

The method of reciprocal monosomics depends upon the existence of suitable monosomic series. These are not universally available. However, it is perhaps not widely known that a large number of monosomic series have been developed. As table 3 indicates, these include varieties produced from most of the major wheat-breeding programmes of the world, as well as many very successful varieties. Saratovskaya 29, for instance, is the major spring wheat of the Soviet Union. Besostaya I had for several years the biggest acreage of any wheat in the world. Caribo and Hobbit are grown widely in Europe. The opportunity therefore exists for screening chromosomal variation among some of the most important wheat germplasm of the world.

Backcross reciprocal monosomics

However, even in the absence of a particular monosomic series, a modified reciprocal monosomic method may be used. This is referred to as the backcross reciprocal monosomic method (Snape & Law 1980), and the details are shown in figure 4. This method requires a further generation of crossing compared with the previous method, and the selection of a number of backcross monosomics to allow for sampling of background variation. The method will, however, permit the detection of major chromosomal differences affecting a quantitative character.

An example of the application of this method is given in table 4. In this experiment, a comparison was made between the estimates of 5A chromosomal effects obtained either by the

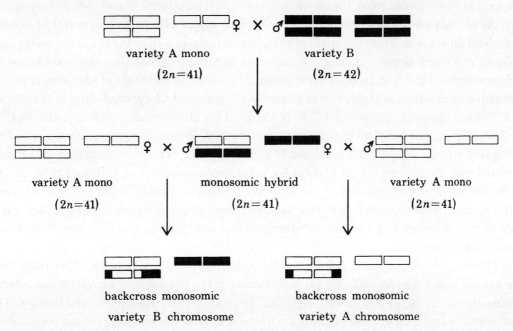

FIGURE 4. The development of backcross reciprocal monosomics by using variety A in which a monosomic series exists and variety B in which monosomics are not available, to give two populations of monosomics with comparable backgrounds but with different hemizygous chromosomes.

TABLE 4. COMPARISON BETWEEN ESTIMATES OF 5A CHROMOSOMAL EFFECTS FROM SUBSTITUTION LINES AND BACKCROSS RECIPROCAL MONOSOMIC (BRM) LINES

	chromosome	height/cm	yield/g
Chinese Spring (CS)	CS5A	107.2	7.01
Chinese Spring (Cap 5A)	Cap5A	99.1	9.89
difference		+8.1*	−2.88*
BRM lines	CS5A	122.7	14.73
BRM lines	Cap5A	114.9	15.15
difference		+7.8*	−0.42
Cappelle-Desprez (Cap)	Cap5A	103.0	15.51
Cappelle-Desprez (Besostaya 5A)	Bes5A	111.2	16.95
difference		−8.2***	−1.44*
BRM lines	Cap5A	109.1	15.16
BRM lines	Bes5A	115.2	21.15
difference		−6.1*	−5.99*

*, $0.05 > p > 0.01$; ***, $p < 0.001$.

backcross reciprocal monosomic method or by inter-varietal chromosome substitutions. The results show that the two sets of estimates are very close to each other. Thus, the effects of chromosome 5A from Chinese Spring compared with its homologue in Cappelle-Desprez and the comparative effects of the 5A chromosomes in Besostaya I and Cappelle-Desprez are very similar for height and, to a lesser extent, yield for both methods.

Direct use of these methods in plant breeding

The aneuploid methods described permit the detection and estimation of chromosomal effects with a precision and generality unattainable by other analytical methods at the moment. However, to what use can these methods be applied in plant breeding? An example of their direct exploitation comes from the work to genetically transform winter wheat into spring wheat. As already mentioned, chromosome 5A has a major effect on the control of winter–spring wheat differences. From the study of substitution lines, as well as the use of the monosomic techniques described already, a range of variation in ear-emergence time has been found for this chromosome (Law *et al.* 1976). At the 'lateness' end of the spectrum of variation is the 5A chromosome from winter wheats such as Cappelle-Desprez and Cheyenne, while at the other is 5A of *Triticum spelta* (a subspecies of *T. aestivum*). This chromosome conferred the smallest vernalization requirement of all the chromosomes studied, so that in converting a winter wheat into a spring wheat, *T. spelta* 5A is obviously the first choice. This chromosome was therefore substituted into the winter wheat Hobbit by using monosomic 5A of this variety as the recurrent parent. After five backcrosses, the Hobbit (*T. spelta* 5A) substitution line was obtained; it behaved exactly as expected and gave ear-emergence times comparable with other spring wheats after a spring sowing. Clearly the substitution of this chromosome had converted Hobbit into a spring wheat.

The substitution of *T. spelta* 5A into Hobbit introduced two major genes other than $Vrn1$. These are the genes for speltoid ears (q) and awning ($b1$). The speltoid character has adverse effects on threshability, so that it was important to remove the q gene by recombination. This was achieved by intercrossing the substitution line and Hobbit. By selecting among the products of this hybrid, or by backcrossing to Hobbit monosomic 5A, four true-breeding types of spring wheat were obtained that were either awned or awnless, or were either speltoid or square-headed. Assessments of the yield produced by these four classes showed variation within and between the four classes, suggesting that *T. spelta* 5A carries separate genes affecting yield not present on Hobbit 5A. The best yields were obtained for lines that were awned and square-headed.

In more extensive yield trials in 1979 and 1980, this spring Hobbit line has outyielded the present-day spring wheats, Highbury and Timmo, by an average of 12% (see table 5).

Table 5. Yield (grams per plot) of Spring Hobbit and two current spring wheat varieties, Timmo and Highbury

	1979	1980	mean	percentage of Highbury
Spring Hobbit	3636	3423	3530	112***
Timmo	3195	3123	3159	100
Highbury	3259	3038	3149	100

***, $p < 0.001$.

It should be stated that these favourable results do not necessarily mean that chromosome manipulation has in this case produced a new commercial variety, because, as might be expected, the Spring Hobbit line carries some of the weaknesses of Hobbit itself. Thus its level of mildew resistance is not sufficient and its bread-making quality is below the level expected for a spring wheat. The Spring Hobbit line must therefore be regarded as a breeders' variety rather than the finished article, and as such it is now being used by spring-wheat breeders as a parent in their breeding programmes.

In the development of Spring Hobbit, a lengthy backcross programme was necessary to substitute *T. spelta* 5A into a Hobbit background. Such a course of action is probably the exception rather than the rule. The more likely approach would be to substitute and fix a desirable chromosome during the early generations of a breeding programme, and then to allow conventional breeding procedures to take their course. In this way some of the genetical deficiencies of Spring Hobbit could have been removed by selection in the development of a new variety.

Conclusions

Intraspecific chromosome manipulation can be expected to contribute directly to breeding programmes under the following circumstances.

1. Where the chromosome has a large effect on a quantitative character and it is not easy to select for this effect by conventional procedures because the means of detection are tedious and labour or space intensive.

2. Where the chromosome has a large effect resulting from the combined effects of many genes that may be difficult to maintain intact by using normal breeding procedures. A special case of this will be the chromosome carrying genes affecting different characters.

The methods of chromosome manipulation may also have value as a screening procedure to detect useful genetic variation among unadapted varieties rarely used by breeders.

Over and above these possibilities of making direct contributions to breeding programmes, chromosome manipulation will continue to have considerable value in carrying out genetical analyses. Ultimately, such analyses should lead to a better understanding of the genetical architecture of wheat. It is from such understanding that opportunities will come for identifying new and useful gene combinations and predicting the outcome of further gene and chromosome manipulation to improving varietal performance.

References (Law *et al.*)

Borlaug, N. E. 1968 Wheat breeding and its impact on world food supply. In *Proc. 3rd Int. Wheat Genetics Symp.*, Canberra, Australia, pp. 1–39.

Brown, M. S. 1966 Attributes of intra- and inter-specific aneuploidy in *Gossypium*. In *Chromosome manipulations in plant genetics* (ed. R. Riley & K. R. Lewis), pp. 98–112.

Gale, M. D. 1978 The effects of Norin 10 dwarfing genes on yield. In *Proc. 5th Int. Wheat Genetics Symp.*, New Delhi, India, pp. 978–987.

Gale, M. D. & Marshall, G. A. 1973 Dwarf wheats and gibberellins. In *Proc. 4th Int. Wheat Genetics Symp.*, Columbia, Missouri, pp. 513–519.

Gale, M. D., Law, C. N. & Worland, A. J. 1975 The chromosomal location of a major dwarfing gene from Norin 10 in new British semi-dwarf wheats. *Heredity, Lond.* **35**, 417–421.

Halloran, G. M. & Boydell, C. W. 1967 Wheat chromosomes with genes for vernalisation response. *Can. J. Genet. Cytol.* **9**, 632–639.

Law, C. N. 1966 The location of genetic factors affecting a quantitative character in wheat. *Genetics, Princeton* **53**, 487–498.

Law, C. N. & Worland, A. J. 1973 Aneuploidy in wheat and its uses in genetic analysis. In *Plant Breeding Institute Annual Report, 1972*, pp. 25–65.

Law, C. N., Scott, P. R., Worland. A. J. & Hollins, T. W. 1976 The inheritance of resistance to eyespot (*Cercosporella herpotrichoides*) in wheat. *Genet. Res.* **25**, 73–79.

Law, C. N., Worland, A. J. & Giorgi, B. 1976 The genetic control of ear-emergence time by chromosome 5A and 5D of wheat. *Heredity, Lond.* **36**, 49–58.

Law, C. N., Young, C. F., Brown, J. W. S., Snape, J. W. & Worland, A. J. 1978 The study of grain protein control in wheat using whole chromosome substitution lines. In *Seed Protein improvement by nuclear techniques*, pp. 483–502. Vienna: I.A.E.A.

Law, C. N., Snape, J. W. & Worland, A. J. 1979 Reciprocal monosomic studies. In *Plant Breeding Institute Annual Report, 1978*, pp. 134–135.

McKewan, J. M. & Kaltsikes, P. J. 1970 Early generation testing as a means of predicting the value of specific chromosome substitutions into common wheat. *Can. J. Genet. Cytol.* **12**, 711–723.

Rajhathy, T. & Thomas, H. 1974 In *Cytogenetics of oats (Misc. Publs Genetics Soc. Canada* no. 2), p. 90.

Sears, E. R. 1953 Nullisomic analysis in common wheat. *Am. Nat.* **87**, 245–252.

Sears, E. R. 1954 The aneuploids of common wheat. *Mo. agric. exp. Stn Res. Bull.* no. **572**, p. 59.

Snape, J. W., Law, C. N. & Worland, A. J. 1976 Chromosome variation for loci controlling ear emergence time on chromosome 5A of wheat. *Heredity, Lond.* **37**, 335–340.

Snape, J. W. & Law, C. N. 1980 The detection of homologous chromosome variation in wheat using backcross reciprocal monosomic lines. *Heredity, Lond.* **45**, 187–200.

Unrau, J. 1950 The use of monosomes and nullisomics in cytogenetical studies of common wheat. *Sci. Agric.* **30**, 66–89.

Interspecific manipulation of chromosomes

By H. Thomas

Welsh Plant Breeding Station, Plas Gogerddan, Aberystwyth, Dyfed SY23 3EB, U.K.

The success of introducing alien variation into crop species from related species depends on the cytogenetic relations between the species. If there are no restrictions on chromosome pairing and recombination in species hybrids, a backcrossing programme can be used to obtain the desired gene transfers. However, when there is a failure of adequate chromosome pairing in species hybrids, techniques of chromosome manipulation have to be used to obtain alien gene transfers. In polyploid crop species it is possible to introduce appropriate single chromosomes of the alien species into the genotype of the recipient species, but the failure of the alien chromosome to become integrated into the genotype of the recipient species often leads to meiotic instability. The introduction of segments of alien chromosomes has been successful through the use of irradiation-induced translocations. The deletions/duplications that are a consequence of such translocations do limit the usefulness of this approach.

In a number of allopolyploid crop species, regular bivalent pairing behaviour has been shown to be genetically controlled. By interfering with the genetic system controlling the diploid-like pairing it is possible to induce pairing between the alien chromosomes and its corresponding chromosomes in the crop species. Gene transfers based on this method involve exchanges between chromosomes of similar gene sequences.

These techniques are discussed and application of the procedures to transfer alien variation into the cultivated oat is described. A scheme is also proposed for transferring the genes controlling regular bivalent pairing from natural polyploid species into synthetic amphiploids in *Lolium/Festuca*.

Introduction

Crop species often lack desirable characters that are found in related species, e.g. resistance to pests and diseases, and it would be advantageous to introduce these traits into the crop species. If the donor and recipient species are closely related and there is no restriction on chromosome pairing in the interspecific hybrid, the transfer of genes from one species to another can be accomplished by conventional plant breeding methods such as backcrossing.

However, conventional plant breeding methods cannot be employed when the chromosomes of the alien species and the crop species are structurally diverged, or there is a genotypic restriction on chromosome pairing in interspecific hybrids. Failure of corresponding chromosomes to pair at meiosis in the interspecific hybrids results in the absence of recombinants in the derivatives of the hybrids. Backcrossing and selecting for the character being transferred in this situation results in the introduction of whole chromosomes as addition or substitution lines (Riley *et al.* 1966; Sharma & Forsberg 1974; Cameron 1966). To introduce desirable alien variation in these circumstances, procedures of chromosome manipulation have to be followed. The techniques of chromosome manipulation used to transfer alien variation into crop species involve the introduction of a segment of the genome of the donor species. In general, such procedures can only be carried out in polyploid species because the genetic duplication inherent in the structure of such species allows the genotype to tolerate some loss and addition of chromosome material.

The greatest success has been achieved in introducing alien variation into bread wheat. The availability of series of aneuploid stocks (Sears 1952) has been the major contributory factor in establishing procedures of gene transfer in wheat. Aneuploid stocks have been widely used to study the genomic structure of bread wheat and its cytogenetic affinity with related species (Sears 1974). An understanding of these cytogenetic relations has led to the establishment of procedures of gene transfer by chromosome manipulation. Partial series of similar aneuploid stocks have been established in cotton (Endrizzi & Ramsay 1979) and oats (Rajhathy & Thomas 1974), and these will facilitate the use of chromosome manipulation in these crop species.

Chromosome addition lines

O'Mara (1940) proposed that an alternative to combining the complete chromosome complements of two species in synthetic amphiploids, as a means of using alien variation, would be to introduce single chromosomes of the donor species into the genotype of the recipient species. O'Mara (1940) was able to add rye chromosomes, with different phenotypic effects, to the wheat genome by backcrossing the amphiploid between the two species to wheat and selecting addition lines from the progeny of the backcrosses. This method of producing single chromosome addition lines has been successfully used to add single chromosomes of alien species to a range of cultivated species (see review by Lacadena 1978).

The effect of the single alien chromosome on the regularity of chromosome behaviour and fertility is variable, depending on which chromosome of the alien species is added to the cultivated species. The addition of some rye chromosomes to wheat disturbed meiotic behaviour, while others had little effect (Riley 1960). Aberrant meiosis leads to instability in the breeding behaviour of the addition lines. This is well illustrated in the disomic addition line combining the pair of chromosomes of the wild oat species *Avena barbata* ($2n = 28$) on which the gene conferring mildew resistance is located with the complement of the cultivated oat *A. sativa* ($2n = 42$) (Thomas *et al.* 1975). The fertility and agronomic performance of the disomic addition line are satisfactory, but the pair of alien chromosomes fails to synapse in 3% of the pollen mother cells. This partial asynapsis leads to the presence of monosomic addition lines ($2n = 43$) in the progeny of the disomic addition line. The transmission of the *A. barbata* monosome is extremely low, and 90% of the progeny of monosomic addition lines revert to the euploid number, and the agronomic attribute (mildew resistance) of the alien chromosome is lost. A large percentage of the disomic addition line reverts to the original mildew susceptible genotype of *A. sativa* in a few generations. Such instability rules out the commercial exploitation of this chromosome addition line.

In general, chromosome addition lines are not suitable for development as commercial varieties on account of their instability, but they form useful material for starting chromosome substitutions and obtaining induced transfers of alien genes.

Chromosome substitutions

The replacement of a pair of chromosomes by a pair of alien chromosomes can be achieved by crossing a plant monosomic for a particular chromosome with a disomic alien addition line. The F_1 hybrid will be monosomic for the relevant chromosome of the recipient species and the alien chromosome. From the progeny of the hybrid it is possible to isolate disomic substitution lines in which the pair of alien chromosomes has replaced the pair of chromosomes of the recipient species. Alternatively the di-monosomic genotype can be crossed with the disomic addition line to give rise to two different genotypes with $21_{II} + 1_I$. One will include 20 pairs plus a monosome of the recipient species and a pair of the alien chromosome. Selfing this plant will yield disomic substitution lines.

If the substitution lines are to be successful the alien chromosome must (1) compensate for the loss of the chromosome of the recipient species, (2) become integrated in the genotype without disturbing meiotic stability and fertility, and (3) confer some agronomic benefit on the crop species.

There are 21×7 possible combinations in wheat–rye substitutions, but only the replacement of genetically similar chromosomes (homoeologous) are effective. The alien chromosome has to compensate for the loss of the genetic activity of the replaced chromosome. The seven chromosomes of rye (Koller & Zeller 1976) and *Agropyron elongatum* (Dvořák 1980) can be classified according to their ability to substitute only for the chromosomes of a single homoeologous group in wheat. The ability of the alien chromosome to compensate for the deletion of homoeologous chromosomes of the recipient species is related to their evolutionary derivation from a common chromosome. Differentiation between the chromosomes that has occurred during speciation will be reflected in the effectiveness with which the chromosomes can replace one another in substitution lines. Dvořák (1980) has reported that the extent of compensation observed was variable when the seven chromosomes of *Agropyron elongatum* replace their homoeologous chromosomes in wheat. Substitutions involving homoeologous group seven were as fertile as Chinese Spring, but homoeologous group two substitutions were nearly sterile. Translocations in the alien and recipient species would alter the genetic correspondence of homoeologous chromosomes and hence their compensating ability.

If substitution lines are to be developed into commercial varieties, the alien chromosome must compensate fully for the loss of the pair of chromosomes and give favourable interactions with the rest of the genotype. Wheat cultivars developed in Germany and eastern Europe have been shown to be substitutions of chromosome 1R of rye for 1B of wheat (Zeller 1973; Mettin *et al.* 1973). Triticale was used as a parent in the breeding of these wheat cultivars and selecting for resistance to disease resulted in the spontaneous establishment of these 1R/1B substitutions. The agronomic performance of these cultivars shows that chromosome 1R of rye compensates fully for 1B of wheat and has become integrated into the wheat genotype.

Introduction of segments of alien chromosomes

The introduction of an alien chromosome either as a substitution or addition line often disturbs the genotypic balance of the recipient species, albeit less than the inclusion of the complete complement in amphiploids. In an attempt to avoid the instability caused by the introduction of a complete alien chromosome, procedures for introducing only segments of the alien chromosomes have been established.

Irradiation-induced translocations

Sears (1956) was the first to use irradiation-induced translocations to transfer resistance to brown rust from *Aegilops umbellulata* to bread wheat. The technique relies on chromosomes rejoining in novel combinations after chromosome breakage induced by the irradiation treatment. Sears (1956) irradiated spikes, just before anthesis, of plants with additions of isochromosomes of *Aegilops umbellulata* carrying a gene for resistance to brown rust, and isolated gene transfers in subsequent progeny. Irradiation of dry seeds of the disomic alien addition can also be successfully used to induce translocations (Driscoll & Jensen 1963). In gene transfers based on irradiation-induced translocations there is a deletion of part of the genome of the recipient species and a duplication arising from the alien segment introduced. The usefulness of the gene transfers will depend on the ability of the genotype to tolerate such deletions and duplications.

TABLE 1. SEGREGATION FOR MILDEW RESISTANCE IN THE F_2 FROM BACKCROSS 4 HYBRIDS

(Figures in parentheses are the percentage of susceptible seedlings.)

recurrent parent	resistant	susceptible
Maris Tabard	157	54 (25.6)
Maris Oberon	185	66 (26.3)
Margam	114	50 (30.5)
Maldwyn	98	47 (32.4)
06618 Cn	131	46 (26.0)

A number of putative translocations, involving the transfer of mildew resistance from the wild oat *Avena barbata* ($2n = 28$) to the cultivated oat *Avena sativa* ($2n = 42$), were identified by Aung (1975) after irradiating seeds of the disomic addition line. One of these lines had normal transmission of the translocation, and the euploid plants, including the translocated chromosome, were normal (Aung *et al.* 1977). Aung & Thomas (1978) described the structure of the translocated chromosome as including most of the long arm of the shortest chromosome of *A. sativa* (ST21 according to Rajhathy's numbering system), and the short arm, which carries the gene for mildew resistance, the centromere and a large segment of the long arm of the *A. barbata* chromosome. Although only a small portion of the *A. sativa* genome is deleted, most of the *barbata* chromosome is included in the translocation. The translocation was originally isolated in the cultivar Manod but it has now been introduced into a further five cultivars of oats.

There is clear evidence that the translocation interacts with the genetic background, both at the gametic and zygotic level. When the backcross 4 hybrids were selfed there was a deficiency of resistant progeny in Margam and Maldwyn backgrounds (table 1). This has been shown to be the result of impaired transmission of pollen including the translocated chromosome in competition with haploid pollen in plants heterozygous for the transfer (Aung & Thomas 1978). The transmission of the translocated chromosome is influenced by the genetic background.

Trials to assess the yield potential of the cultivars including the translocation compared with the recipient cultivars have been undertaken by I. T. Jones (personal communication) at the Welsh Plant Breeding Station. Since the original cultivars were susceptible to mildew, they were sprayed with fungicide to eliminate the effect of mildew infection on yield. The advantage of the resistance to mildew conferred by the transfer is nullified by spraying, and it is possible to assess the effect of the translocation on yield in different genotypic backgrounds by comparing the performance of the transfers with the original cultivar. In table 2 the yield of the translocation lines (BC$_4$) are expressed as the yield of corresponding cultivars. It is clear that the presence of the translocation interacted with the genetic backgrounds. In Maris Tabard the yield of the transfer was equal to the original cultivar while the presence of the translocation significantly depressed yield in Margam. The highest yielding cultivar was the W.P.B.S. selection 06618 Cn and the yield of the transfer was only 3.21% less than the original cultivar. These results show the importance of assessing such transfers in as wide a range of genetic background as possible in a breeding programme. Intercrossing the different cultivars, which are homozygous for the translocations, will give further opportunity for selecting more favourable genetic backgrounds without having to select for resistance to mildew. The advantage of the

TABLE 2. YIELD OF TRANSLOCATION LINES IN DIFFERENT GENETIC BACKGROUNDS (BACKCROSS 4)

recurrent spray	yield of translocation line as percentage of original cultivar sprayed	as percentage of original cultivar unsprayed
Maldwyn	93.48	102.70
Margam	78.18	83.41
Maris Tabard	100.85	108.38
06618 Cn	96.79	102.48
Maris Oberon	91.12	94.75

incorporation of the mildew resistance is shown in the comparison between the yield of the transfers and the original cultivars when the latter were not sprayed (table 2).

Knott (1968) and Dvořák & Knott (1977) have demonstrated that a number of transfers of *Agropyron* genes into wheat based on induced translocations are exchanges between homoeologous chromosomes. Selection for regular transmission has led to the isolation of transfers in which the deletion of the recipient wheat chromosome is compensated for by a segment of the homoeologous *Agropyron* chromosome. In the *Avena* transfer, the *A. barbata* and *A. sativa* chromosomes involved in the translocation were not homoeologous (Aung & Thomas 1978). Transec, a wheat-rye irradiation-induced translocation, has also been shown to be an exchange between non-homoeologous chromosomes by Driscoll & Anderson (1967). In these translocations the loss of a small portion of the genome of the recipient species is tolerated.

GENETIC-INDUCED GENE TRANSFERS

The failure of alien chromosomes to pair with their corresponding chromosomes in the cultivated species is not always associated with structural differentiation of the chromosomes but with the genetic control of chromosome pairing. In wheat, regular bivalent pairing is controlled by the *Ph* gene on the long arm of chromosome 5B (Riley & Chapman 1958; Sears & Okomota 1958). The *Ph* gene restricts pairing to homologous chromosomes and the homoeologues of the three constituent genomes synapse and form chiasmata associations only when chromosome 5B is deleted. The presence of such a genetic system in natural allopolyploids is also responsible for minimizing pairing between alien chromosomes and their homoeologues in the crop species. The relaxation of the control mechanism leads to pairing between alien chromosomes and the corresponding wheat chromosomes (Riley & Chapman 1963; Dvořák & Knott 1972). By manipulating the control mechanism to allow homoeologous chromosomes to pair, Riley *et al.* (1968) and Sears (1973, 1978) have been able to induce alien chromosomes to pair with their homoeologues in wheat and isolate recombinants including genes located on the alien chromosome. Riley *et al.* (1968) used genotypes of *Aegilops speltoides*, which suppresses the activity of the *Ph* gene, while Sears (1973), using the appropriate aneuploid lines, deleted chromosome 5B to induce pairing between alien chromosomes and their homoeologues in wheat.

TABLE 3. THE CROSSING SCHEME USED TO ISOLATE GENETICALLY INDUCED TRANSFER OF MILDEW RESISTANCE FROM *A. BARBATA* TO THE CULTIVATED OAT

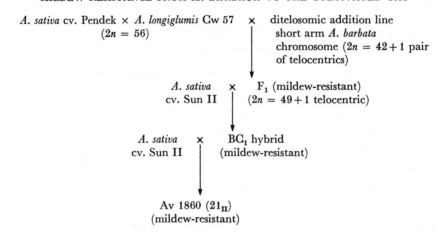

The techniques were developed by wheat cytogeneticists based on their understanding of the genomic structure and meiotic behaviour of wheat and its related species. There is strong evidence that regular bivalent pairing in *Avena sativa* is also genetically controlled, although the precise chromosome or chromosomes carrying the effective genes have not been definitely identified (Rajhathy & Thomas 1974). A genotype of the diploid species *A. longiglumis* (Cw 57) has been reported to modify the genetic mechanism controlling bivalent pairing in *A. sativa* and induce homoeologous chromosome pairing (Rajhathy & Thomas 1972; Thomas & Al-Ansari 1980) in a manner comparable to *Aegilops speltoides* in wheat.

The effect of the *A. longiglumis* (Cw 57) has been successfully used to transfer mildew resistance from *A. barbata* to the cultivated oat (Thomas *et al.* 1980). The crossing scheme is represented in table 3. The identification of progeny resistant to mildew (Av 1860), in which the telocentric is absent, is proof of crossing over between the *A. barbata* and *A. sativa* chromosomes. In hybrids between *A. sativa* and Av 1860, 21 bivalents were regularly formed. In the absence of the Cw 57 genotype, no natural recombinants were recovered from the progeny of the monosomic addition line (Thomas *et al.* 1975). The isolation of recombinants is attributable to the induced pairing of homoeologous chromosomes in the presence of the Cw 57 genotype.

In contrast to irradiation-induced translocations, all exchanges in genetic-induced transfers are between homoeologous chromosomes. The introduced segment of the alien chromosome replaces a segment of a genetically equivalent chromosome of the recipient species. In view of the evolutionary derivation of the chromosomes from a common chromosome, the segment of the alien chromosome has a good chance of compensating fully for the segment of the chromosome of the recipient species replaced. The degree of compensation will depend on how close the gene order has been conserved in the alien and homoeologous chromosomes during evolution. Sears (1973, 1978) has reported variation in the transmission of a number of wheat–*Agropyron* transfers, which reflect such differences.

When the introduction of alien genes is dependent on recombination as a result of the suppression of the control mechanism, there are better opportunities for restricting the size of the segment of the alien chromosome incorporated. In the *Agropyron* transfers described by Sears (1978) there was variation in the size of the segment of the alien chromosome introduced. The ability to isolate transfers based on smaller segments of the alien chromosome improves the chances of breaking undesirable linkages found in whole chromosome substitution and additions. Genetically induced chromosome pairing introduces greater precision as a method of gene transfer than irradiation-induced translocations. However, irradiation remains a valuable technique when the alien chromosomes fail to synapse with related chromosomes in the crop species on account of structural differentiation.

In situations where the alien chromosome fails to pair even when homoeologous pairing is induced, Sears (1972) proposed a method based on fusion of telocentric arms. In plants monosomic for an alien and a recipient species chromosome, misdivision of the univalents and union of the resulting telocentric would give rise to a chromosome including an arm from both univalents. If the two chromosomes concerned were homoeologous there would be a good chance of compensation, but in the event of fusion between arms of non-homoeologous chromosomes, success would depend on the ability of the genotype to tolerate the deletion and duplication. Sears (1972) has produced wheat–rye translocations by using this method.

Procedures to transfer alien variation into crop species have been developed to overcome barriers that have arisen during speciation to prevent gene flow and recombination. Stable meiotic behaviour in natural allopolyploids is ensured by the restriction of pairing to homologous chromosomes. The presence of such genetic systems has maintained the integrity of the constituent genomes during evolution. The development of commercial varieties based on synthetic amphiploids combining diploid genomes has often been hampered by irregular meiotic behaviour including allosyndetic pairing, which leads to the dissipation of advantageous gene combinations. If the genetic system in natural allopolyploids could be transferred into synthetic polyploids it should stabilize the meiotic behaviour of the newly synthesized allopolyploids.

The natural allopolyploid *Festuca arundinacea* ($2n = 6x = 42$) has regular bivalent pairing and disomic inheritance (Lewis *et al.* 1980), two features of homogenetic pairing in allopolyploids. Jauhar (1975) proposed that the restriction of chromosome pairing to homologous pairs is controlled by a gene or genes that are ineffective in the hemizygous conditions. Data on chromosome pairing between natural bivalent-pairing tetraploid and hexaploid species of *Festuca* produced by M. Borrill and W. G. Morgan at this station also show evidence of the genetic control of chromosome pairing in these polyploid species. In euhaploids and interspecific hybrids with diploid species of *Festuca* and *Lolium*, the control system does not operate and allosyndetic pairing occurs freely. It should therefore be possible to isolate recombinants, including the *F. arundinacea* pairing genes in *Lolium* genotypes. We are now attempting to establish such gene transfers at this station and use them to improve the regularity of meiotic behaviour and the stability of tetraploid hybrid combinations involving *Lolium* species. Since there is a tendency to preferential pairing in *L. perenne* × *L. multiflorum* synthetic tetraploids (Lewis 1980), the inclusion of the pairing genes in the disomic condition should further enhance preferential pairing and the stability of the amphiploids.

Further refinements of techniques of chromosome manipulation for the transfer of alien genes into crop species will probably be concerned with reducing the size of the segment of the alien chromosome introduced in such transfers and consequently the chances of introducing any undesirable linked genes. Pandey (1978, 1980) has described the introduction of single genes in *Nicotiana* by using pollen that had been treated with a large dose of ionizing radiation. Segments of DNA from the treated pollen are incorporated into the egg of the recipient species and the 'genetically transformed' egg developed parthenogenetically into a viable seed. More information is required on the mechanisms involved in the transformation of the egg and the doubling of the chromosome number of the egg leading to normal seed development before the wider application of the technique can be evaluated.

REFERENCES (Thomas)

Aung, T. 1975 Studies of the use of irradiation in the incorporation of alien variation in *Avena*. M.Sc. thesis, University of Wales.

Aung, T. & Thomas, H. 1978 The structure and breeding behaviour of a translocation involving the transfer of mildew resistance from *Avena barbata* into the cultivated oat. *Euphytica* 27, 731–739.

Aung, T., Thomas, H. & Jones, I. T. 1977 The transfer of the gene for mildew resistance from *Avena barbata* (4x) into the cultivated oat *A. sativa* by an induced translocation. *Euphytica* 26, 623–632.

Cameron, D. R. 1966 Some experimental applications of aneuploidy in *Nicotiana*. In *Chromosome manipulations and plant genetics* (ed. R. Riley & K. R. Lewis), pp. 1–7. Edinburgh: Oliver & Boyd.

Driscoll, C. J. & Anderson, L. M. 1967 Cytogenetic studies of Transec – a wheat rye translocation line. *Can. J. Genet. Cytol.* 9, 375–380.

Driscoll, C. J. & Jensen, N. F. 1963 A genetic method for detecting induced intergeneric translocations. *Genetics* 48, 459–468.

Dvořák, J. 1977 Homoeologous chromatin transfer in a radiation-induced gene transfer. *Can. J. Genet. Cytol.* 19, 125–131.

Dvořák, J. 1980 Homoeology between *Agropyron elongatum* chromosomes and *Triticum aestivum* chromosomes. *Can. J. Genet. Cytol.* 22, 237–259.

Dvořák, J. & Knott, D. R. 1972 Homoeologous pairing of specific *Agropyron elongatum* (2n = 70) chromosome with wheat chromosomes. *Wheat Inf. Serv. Kyoto Univ.*, nos. 33 and 34, pp. 35–37.

Endrizzi, J. E. & Ramsay, G. 1979 Monosomes and telosomes for 18 of the 26 chromosomes of *Gossypium hirsutum*. *Can. J. Genet. Cytol.* 21, 531–536.

Jauhar, P. 1975 Genetic control of diploid-like meiosis in hexaploid tall fescue. *Nature, Lond.* 254, 595–597.

Knott, D. R. 1968 Translocations involving *Triticum* chromosomes and *Agropyron* chromosomes carrying rust resistance. *Can. J. Genet. Cytol.* 10, 695–696.

Koller, O. L. & Zeller, F. J. 1976 The homoeologous relationships of rye chromosomes 4R and 7R with wheat chromosomes. *Genet. Res.* 28, 177–188.

Lacadena, J. R. 1978 Interspecific gene transfer in plant breeding. In *Interspecific hybridization in plant breeding* (*Proc. 8th Congress of Eucarpia*, Madrid, 1977), pp. 45–62.

Lewis, E. J. 1980 Chromosome pairing in tetraploid hybrids between *Lolium perenne* and *L. multiflorum*. *Theor. appl. Genet.* 58, 137–143.

Lewis, E. J., Humphreys, M. W. & Caton, M. P. 1980 Disomic inheritance in *Festuca arundinacea* Schreb. *Z. PflZücht.* 84, 335–341.

Mettin, W. D., Bluthner, W. D. & Schlegel, C. 1973 Additional evidence on spontaneous 1B/1R wheat/rye substitutions and translocations. In *Proc. 4th Int. Wheat Genet. Symp.*, Missouri Agric. Expl. Stn, Columbia, pp. 179–184.

O'Mara, J. G. 1940 Cytogenetic studies on Triticale. 1. A method for determining the effect of individual *Secale* chromosomes on *Triticum*. *Genetics, Princeton* 25, 401–408.

Pandey, K. K. 1978 Gametic gene transfer by means of irradiated pollen. *Genetica* 49, 53–69.

Pandey, K. K. 1980 Further evidence for egg transformation in *Nicotiana*. *Heredity, Lond.* 45, 15–29.

Rajhathy, T. & Thomas, H. 1972 Genetic control of chromosome pairing in hexaploid oats. *Nature, new Biol.* 239, 217–219.

Rajhathy, T. & Thomas, H. 1974 Cytogenetics of oats (*Avena* L.). *Misc. Pub. genet. Soc. Can.* no. 2. Ontario: The Genetics Society of Canada.

Riley, R. 1960 The meiotic behaviour, fertility and stability of wheat-rye chromosome addition lines. *Heredity, Lond.* 14, 89–100.

Riley, R. & Chapman, V. 1958 Genetic control of the cytologically diploid behaviour of hexaploid wheat. *Nature, Lond.* **182**, 713–715.
Riley, R. & Chapman, V. 1963 The effects of the deficiency of chromosome V (5B) of *Triticum aestivum* on the meiosis of synthetic amphiploids. *Heredity, Lond.* **18**, 473–484.
Riley, R., Chapman, V. & Johnson, R. 1968 The incorporation of alien disease resistance in wheat by genetic interference with the regulation of meiotic chromosome synapsis. *Genet. Res.* **12**, 199–219.
Riley, R., Chapman, V. & Macer, R. C. F. 1966 The homoeology of an *Aegilops* chromosome causing stripe rust resistance. *Can. J. Genet. Cytol.* **88**, 616–630.
Sears, E. R. 1952 The aneuploids of common wheat. *Mo. agric. exp. Stn Res. Bull.* no. **572**, pp. 1–59.
Sears, E. R. 1956 The transfer of leaf-rust resistance from *Aegilops umbellulata* to wheat. *Brookhaven Symp. Biol.* **9**, 1–22.
Sears, E. R. 1972 Chromosome engineering in wheat. In *Stadler Symposia*, vol. 4, pp. 23–38.
Sears, E. R. 1973 *Agropyron* – wheat transfers induced by homoeologous pairing. In *Proc. 4th Int. Wheat Genet. Symp.*, Missouri Agric. Expl. Stn, Columbia, pp. 191–200.
Sears, R. R. 1974 The wheats and their relatives. In *Handbook of genetics*, vol. 2 (*Plants, plant viruses and protists*) (ed. R. C. King), pp. 59–91. New York: Plenum.
Sears, E. R. 1978 Analysis of wheat–*Agropyron* recombinant wheat chromosomes. In *Interspecific hybrids in plant breeding* (*Proc. 8th Congress of Eucarpia*, Madrid, 1977), pp. 63–72.
Sears, E. R. & Okamoto, M. 1958 Intergenomic chromosome relationships in hexaploid wheat. In *Proc. X Int. Congr. Genet.*, vol. 2, pp. 258–259.
Sharma, D. C. & Forsberg, R. A. 1974 Alien chromosome substitution – a cause of instability for leaf rust resistance in oats. *Crop Sci.* **14**, 533–536.
Thomas, H. & Al-Ansari, N. 1980 Genotypic control of chromosome pairing in *Avena longiglumis* × *A. sativa* hybrids. *Chromosoma* **79**, 115–124.
Thomas, H., Leggett, J. M. & Jones, I. T. 1975 The addition of a pair of chromosomes of the wild oat *Avena barbata* (2n = 28) to the cultivated oat *A. sativa* L. (2n = 42). *Euphytica* **24**, 717–724.
Thomas, H., Powell, W. & Aung, T. 1980 Interfering with regular meiotic behaviour in *Avena sativa* as a method of incorporating the gene for mildew resistance from *A. barbata*. *Euphytica* **29**. (In the press.)
Zeller, F. J. 1973 1B/1R Wheat–rye chromosome substitutions and translocations. In *Proc. 4th Int. Wheat Genet. Symp.*, Missouri Agric. Expl. Stn, Columbia, pp. 209–222.

The control of recombination

By R. Riley, F.R.S., C. N. Law and V. Chapman

Plant Breeding Institute, Maris Lane, Trumpington, Cambridge CB2 2LQ, U.K.

Recombination results in the release of variation upon which selection is practised in plant breeding. However, according to the objectives of the programme it may be necessary to reduce recombination to limit the disturbance of arrangements of genes already well suited to agricultural needs or to increase recombination to maximize the likelihood of recombination's giving rise to transgressive segregation. Appropriate breeding manipulations are discussed. In addition, descriptions are provided of the induction of recombination between chromosomes that are distantly related in evolution and between which meiotic pairing and recombination does not normally take place.

Since most plant breeding is dependent upon the release of variation as a consequence of recombination and segregation, a good deal of attention has been directed in plant breeding research to the management of this variation and to the control of its release. From the outset it is important to establish that the efficiency of the breeding process may be improved on some occasions by increased, and on other occasions by reduced, recombination. The more usual condition is that genetic advance can only be achieved by maximizing the range of variation in the population concerned, to provide transgressive segregation among the extremes of which selection can be practised. Alternatively, genetic advance may be of the most practical use when already well adjusted genotypes are disturbed to the minimum extent necessary to incorporate new, perhaps simply inherited, beneficial characters. It was to satisfy the latter objective that the backcross system of breeding was developed. It is appropriate to make this point since many of the systems of breeding used in variety production, like backcrossing, have the purpose of adjusting the consequence of recombination. In the present discussion, by contrast, attention will be concentrated on the direct control of recombination rather than on adjusting its effects. More attention has been paid to the control of recombination in wheat than in other genera, so that most of our examples will be drawn from the wheat crop group.

Relation of recombination to hybridity

The conventional processes of variety production in wheat (*Triticum aestivum*) are of classical pedigree selection from intervarietal hybrids, or some variant of this involving bulk selection, or the isolation of individual lines from heterogeneous populations generated from intervarietal hybrids. In addition, backcrossing is important where characters of complex inheritance, such as grain quality, must be preserved as the first requirement of successful varieties. Curiously, chiasma frequency – as represented by the frequency of bivalent chromosomes at the first metaphase of meiosis – is lower in intervarietal F_1s than in their varietal parents. By diallel cross analysis, using data obtained by Watanabe (1962), Riley & Law (1965) showed that this negative heterosis for chromosome pairing was largely the product of non-allelic gene interaction. Person (1956) showed in the backcross derivatives of intervarietal wheat hybrids that there was a higher frequency of chromosome pairing with an increase in the expected homozygosity. However, the return to the parental level of chiasma frequency was delayed relative to the return to homozygosity.

This example of the control of chromosome pairing, which shows a coarse association with the extent of heterozygosity, implies that the maintenance of linkage and of synteny may be high in intervarietal wheat hybrids and that, if a high frequency of recombination is sought, it may be necessary to prolong the period of maximum heterozygosity by intercrossing early generation segregants or by top-crossing. However, intercrossing F_2 plants derived from hybrids between two west European winter wheat varieties led to no greater release of variation because of the insignificance of linkage for the characters studied, which included height and grain yield (Snape 1979). The delayed return, following heterozygosity, to the maximum level of chromosome pairing may cause the production of aneuploids precisely at that stage at which new varieties are being multiplied for release, when genetic instability often creates problems for breeders. If a high standard of stabilization is demanded, there is no resolution to this problem that does not either delay the availability of improved genotypes to agriculture or diminish, by the introduction of backcrosses, the range of genetic variability among which the breeder can select.

So far reference has been made to aspects of the genetic control of chromosome pairing, and hence recombination, that must inevitably be considered by the wheat breeder and which can be manipulated according to the objective of the programme. We shall now describe breeding manipulation that can be used to alter recombination.

Limitation of recombination
Haploids

In inbreeding crop species, such as wheat and barley, the continuation of recombination and segregation in a number of the generations derived from intervarietal hybrids delays the fixation in the homozygous state of new potential varieties. Ways of circumventing this delay and of reducing the extent of recombination are provided by haploid breeding methods. In their simplest form these consist of deriving haploid sporophytes directly from the gametophytes produced by F_1 hybrids. Such sporophytes can be treated with substances, such as colchicine, which inhibit spindle formation and so have the chromosome numbers of some of their cells returned to the diploid condition. In these diploid sectors and in the lines derived from them the first products of recombination and segregation from the F_1 are fixed and, in inbreeding crops, each is potentially a new variety.

Much attention has been paid to the feasibility of culturing anthers or pollen grains in nutrient media and stimulating the development of haploid sporophytes from the pollen grains. This is a procedure that has proved particularly effective in the Solanaceae (Nitzsche & Wenzel 1977) but less so in the Leguminosae and the Gramineae. However, in the latter, for barley and possibly for wheat, there are alternative opportunities for haploid production. These arise because in F_1 hybrids between *Hordeum vulgare* (cultivated barley) and *Hordeum bulbosum* (a wild perennial) the chromosomes of *H. bulbosum* are frequently eliminated in the first few cell divisions in the embryo, leaving a sporophyte with only the haploid set of chromosomes of *H. vulgare* (Kasha & Kao 1970). The phenomenon of chromosome elimination permits the use of haploid breeding methods and the fixation of the first products of F_1 meiosis in barley breeding and thus the limitation of the breeding cycle to one occasion for recombination.

Subsequently, Barclay (1975) discovered that F_1 hybrids from the cross of *T. aestivum* × *H. bulbosum* developed into wheat haploids. Unfortunately the crossability of wheat with *H. bulbosum* proved to be confined, probably because of the activity of alleles at two loci, to some of varieties of Asian and Australian origin. There is as yet, therefore, no generally available methodology for the limitation of recombination in the derivatives of intervarietal F_1s in wheat.

Recombination confined to a single chromosome pair

Cytogenetic methods in wheat by which recombination could be confined to a single chromosome pair were first used by Law (1966). The procedure depends upon the initial construction of chromosome substitution lines in which a single unmodified pair of chromosomes from a donor variety replaces its homologous pair in a recipient variety. The background chromosomes in such a substitution line are those of the recipient variety, which are unchanged.

Hybridization between the recipient variety and the substitution line produces a hybrid in which there is heterozygosity for genes only in the pair with a substituted chromosome, while the background chromosomes are entirely homozygous. When this single chromosome heterozygote is used to pollinate a plant which is monosomic for the heterozygous chromosome, monosomic derivatives can be selected. In this case, the monosomic chromosome is that which was heterozygous in the pollen parent, so the first products of recombination and segregation are fixed in the hemizygous condition. Selfing of this recombinant-chromosome hemizygote will lead to the fixation of the chromosome in the disomic condition in euploid plants. Opportunities are thus offered of selecting for precise recombination in a single chromosome and thus for unusually 'fine tuning' of a plant genotype.

CONTROL OF HOMOEOLOGOUS RECOMBINATION

In euploids of *T. aestivum* ($2n = 6x = 42$) genetically corresponding (homoeologous) chromosomes of the different sets assembled at each advance in polyploidy do not pair at meiosis, principally because of the activity of the allele at the *ph* locus on chromosome 5B (for key to references see Sears 1977). In mutants of *ph* or when chromosome 5B is deficient, or when alien dominant or epistatic alleles are present, homoeologous chromosomes will pair and recombine (for references, see Riley (1974)).

In hybrids between *T. aestivum* and many of its relatives, in the genera *Aegilops*, *Secale*, *Agropyron* and *Haynaldia*, there is little or no pairing at meiosis between the chromosomes of the parental species. Consequently there is little or no recombination between wheat chromosomes and those of such relatives so that, in breeding, their potentially useful genes cannot be transferred to wheat chromosomes. The relation between the chromosomes of wheat and those of such relatives is one of homoeology, so that removal of pairing suppression by the *ph* activity results in pairing and recombination. Experimental interference with the *ph* activity in interspecific hybrids thus provides access to a new range of genetic variation.

The first example of the transference of alien genetic variation to wheat by induced homoeologous recombination involved the incorporation in *T. aestivum* of rust resistance from *Aegilops comosa* (Riley *et al.* 1968). In this work, the single chromosome (2M) of *Ae. comosa* that determines stripe rust resistance, and which would not normally pair with its wheat homoeologues, was isolated, by backcrossing, from all other *Ae. comosa* chromosomes in plants that also had the full complement of wheat chromosomes. Such monosomic 2M addition plants were then pollinated by *Aegilops speltoides* to produce hybrids in which, because *Ae. speltoides* has a genetic activity dominant or epistatic to *ph*, there was homoeologous chromosome pairing at meiosis and chromosome 2M could pair and recombine with its wheat homoeologues. Backcrossing to wheat followed and this resulted in the isolation of phenotypically normal euploid forms of *T. aestivum* with the rust resistance of *Ae. comosa* in which recombination had occurred between chromosome 2M and 2D. The recombinant chromosome possessed the short arm, the centromere and the proximal part of the long arm of 2M and the distal part of the right arm of 2D.

Subsequently a line was isolated with the segment of chromosome 2M determining rust resistance incorporated by recombination into chromosome 2A. This line arose in the same segregating population in which 2M–2D recombination had occurred.

Recombination between 2A and 2M was also obtained by using plants that were simultaneously nullisomic for 5B and tetrasomic for 5D. Because 5B is absent there is homoeologous meiotic chromosome pairing, but other effects of the deficiency of 5B are largely corrected by the compensating effects of tetrasomy for 5D. A crossing programme was followed that created a line nullisomic for 5B, tetrasomic for 5D and monosomic for 2M. Homoeologous pairing at meiosis enabled recombination to occur between 2M and its wheat homoeologues. After this cytogenetic condition had been maintained for one or more meiotic cycles, the nulli-5B, tetra-5D, mono-2M plants were pollinated by wheat plants tetrasomic for chromosome 5B. The derivatives of the cross had chromosome 5B returned to the disomic state, were trisomic for 5D and may or may not have carried chromosome 2M. In the next generation, which was obtained by selfing, selection was practised for 42-chromosome euploid plants with rust resistance. These segregants had undergone homoeologous recombination. The structure of the 2A/2M recombinant chromosome was analogous to that of the 2D/2M chromosome derived by the *Ae. speltoides ph*-suppression procedure. Methods corresponding to these were also used by Sears (1973) to induce homoeologous recombination between wheat and *Agropyron* chromosomes to incorporate alien leaf rust resistance in wheat.

Promising work is being carried out by P. Harris of the Plant Breeding Institute, on the transference from chromosome $1C^u$ of *Ae. umbellulata* to wheat of genes controlling high molecular mass storage glutenin proteins of the grain, which may be important in bread-making. The process that has been used to induce homoeologous recombination is based on the creation of nullisomy for 5B and tetrasomy for 5D in the presence of $1C^u$ but, since $1C^u$ was introduced from a line in which it was substituted for wheat chromosome 1A, it may be present as a monosomic addition or a monosomic substitution. It is suspected that 42-chromosome homoeologous recombinants have been isolated after the restoration of 5B disomy by the cross with tetrasomic 5B, because some plants possess the high molecular mass glutenins that mark chromosome $1C^u$ but not the gliadin bands that are also determined by genes on the same chromosome.

There is consequently ample evidence of manipulation of the *ph* system leading to the ready incorporation in wheat by recombination of simply inherited genetic variation of related species especially in the genera *Aegilops* and *Agropyron*. A surprising anomaly is that the frequency of meiotic pairing and recombination is very low between wheat and rye chromosomes, even when the *ph* activity does not occur. This is so even though there is much evidence to indicate the close genetic relationship between wheat and rye, such as the ability of rye chromosomes to compensate genetically for their wheat homoeologues (Riley 1965) and the high frequency of DNA segments with closely comparable nucleotide sequences (Flavell *et al.* 1977). There is some evidence of meiotic pairing between wheat and rye chromosomes (Bielig & Driscoll 1970; Mettin *et al.* 1976), but it apparently takes place with great rarity and there is no clear evidence of crossing-over.

Something of the nature of the genetic control of homoeologous meiotic pairing in rye was described by Riley *et al.* (1973). A genetic activity of the short arm of rye chromosome 5R is responsible for the increase of homoeologous pairing in wheat–rye hybrids. However, it is uncertain whether in plants with the highest level of pairing which had the haploid set of 21 chromosomes of *T. aestivum* (namely the ABD genomes) and two complete genomes of rye (RR) and additionally the short arm of $5R^s$, had any wheat–rye recombination. Probably this could only be detected by breeding experiments, and such have yet to be successfully undertaken with this highly infertile material.

Nevertheless, the derivation of recombinants between wheat and rye, which enables the association of the beneficial characters of both crops without the inadequacies of triticale, remains an important objective in the exploitation of the control of recombination. It is tempting to speculate that the failure of wheat–rye pairing is related to the presence of telomeric heterochromatin that occurs in the chromosomes of *S. cereale* and not in those of wheat and which may represent up to 18% of the nuclear DNA content of rye (Bedbrook *et al.* 1980). If meiotic synapsis is initiated terminally or the positioning of chromosomes before synapsis depends upon the composition of telomeric DNA, then modification of the distinctive structure of rye chromosomes may be a necessary preliminary to the occurrence of recombination between homoeologous chromosomes of wheat and rye.

Conclusions

The genetic control of recombination will influence the course of plant breeding programmes whether or not breeders attempt to intervene in the process. Clearly the structure of any programme will make allowance for this as one of the factors bearing upon the likelihood of success and the time course of the work. It is also apparent that the breeder can intervene in the control of recombination and obtain recombinants that would not have occurred without such intervention. These possibilities have arisen as a result of the availability of knowledge of the genetic and cytogenetic systems through which control is exercised. As further knowledge is provided of these systems, and particularly of such components of them as may be represented by telomeric structure in rye, intervention in the control of recombination will have greater precision than is now possible.

References (Riley *et al.*)

Barclay, I. R. 1975 High frequencies of haploid production in wheat (*Triticum aestivum*) by chromosome elimination. *Nature, Lond.* **256**, 410–411.

Bedbrook, J. R., Jones, J., O'Dell, M., Thompson, R. D. & Flavell, R. B. 1980 A molecular description of telomeric heterochromatin in *Secale* species. *Cell* **19**, 545–560.

Bielig, L. & Driscoll, C. J. 1970 Substitution of rye chromosome $5R^n$ for chromosome 5B of wheat and its effect on chromosome pairing. *Genetics, Princeton* **65**, 241–247.

Flavell, R. B., Rimpau, J. & Smith, D. B. 1977 Repeated sequence DNA relationships in four cereal genomes. *Chromosoma* **63**, 205–222.

Kasha, K. J. & Kao, K. N. 1970 High frequency haploid production in barley (*Hordeum vulgare* L.) *Nature, Lond.* **225**, 874–876.

Law, C. N. 1966 The location of genetic factors affecting a quantitative character in wheat. *Genetics, Princeton* **53**, 487–498.

Mettin, D., Schlegel, R., Blüthner, W. D. & Weinrich, M. 1976 Giemsa-banding von M1-Chromosomen bei Weizen-Roggen-Bastarden. *Biol. Zbl.* **95**, 35–41.

Nitzsche, W. & Wenzel, C. 1977 Haploids in plant breeding. *Fortschr. PflZücht.* **8**, 1–101.

Person, C. 1956 Some aspects of monosomic wheat breeding. *Can. J. Bot.* **34**, 60–70.

Riley, R. 1965 Cytogenetics and wheat breeding. In *Genetics Today, Proceedings of the 11th International Congress of Genetics* (ed. S. J. Geerts), **3**, 681–688. Oxford: Pergamon Press.

Riley, R. 1974 Cytogenetics of chromosome pairing in wheat. *Genetics, Princeton* **78**, 193–203.

Riley, R., Chapman, V. & Johnson, R. 1968 The incorporation of alien disease resistance in wheat by genetic interference with the regulation of meiotic chromosome synapsis. *Genet. Res.* **12**, 199–219.

Riley, R., Chapman, V. & Miller, T. E. 1973 The determination of meiotic chromosome pairing In *Proceedings of the 4th International Wheat Genetics Symposium*, Columbia, Missouri (ed. E. R. Sears & L. M. S. Sears), pp. 731–732. Agricultural Experiment Station, University of Missouri.

Riley, R. & Law, C. N. 1965 Genetic variation in chromosome pairing. *Adv. Genet.* **13**, 57–114.

Sears, E. R. 1973 *Agropyron*-wheat transfers induced by homoeologous pairing. In *Proceedings of the 4th International Wheat Genetics Symposium*, Columbia, Missouri (ed. E. R. Sears & L. M. S. Sears), pp. 191–199. Agricultural Experiment Station, University of Missouri.

Sears, E. R. 1977 An induced mutant with homoeologous pairing in common wheat. *Can. J. Genet. Cytol.* **19**, 585–593.

Snape, J. W. 1979 Detection of linkage for quantitative characters in wheat. In *Proceedings of 5th International Wheat Genetics Symposium*, New Delhi (ed. S. Ramanujam), pp. 591–596. New Delhi: Indian Society of Genetics and Plant Breeding.

Watanabe, Y. 1962 Meiotic irregularities in intervarietal hybrids of common wheat. *Wheat Inform. Serv.* **14**, 5–7.

Discussion

R. JOHNSON (*Plant Breeding Institute, Cambridge, U.K.*). Dr Riley showed a graph at the Meeting, indicating that pairing of chromosomes was reduced in F_1 hybrids and gradually reverted to normal over subsequent generations. However, restoration of full pairing appeared to lag behind the return to full homozygosity. Has chromosome pairing been examined in doubled haploids produced from F_1 hybrids or from segregating material in subsequent generations?

R. RILEY. As far as I am aware, no detailed study of chromosome pairing of derived double haploids obtained from F_1 wheat hybrids has been carried out. Undoubtedly such an investigation would be of interest and might provide useful information about the relations between particular homozygotes, chromosome pairing and stability.

Perspectives in chromosome manipulation

By C. J. Driscoll

Department of Agronomy, Waite Agricultural Research Institute, The University of Adelaide, South Australia, Australia 5064

Three categories of chromosome manipulation are discussed, with examples from hexaploid wheat. First, uncontrolled events, such as intergeneric translocations induced by mutagenic agents, have been frequently isolated but have been infrequently incorporated into widely grown varieties. A method proposed by K. W. Shepherd is designed to select an accommodating genetic background for these interchanges and thereby overcome this deficiency. This relies on selection for yield among many selections that are homozygous for a translocation but segregating for many other genetic components.

The second category involves manipulations associated with the distinctive genetic activity of chromosome 5B. *ph* mutants are expected to play an important role in the transfer of genetic material from other genera to wheat. A method described by E. R. Sears is designed to isolate a small intercalated alien segment. This relies on crossing-over in an alien segment that is common to two distinct types of homoeologous-exchange chromosomes.

The third category involves manipulations associated with a male-sterility mutation on chromosome 4A and the distinctive genetic activities of this chromosome. A method, described by the author, is designed to produce hybrid wheat with an induced male-sterility mutation on this chromosome. Fertility restoration is being attempted with four chromosomes, a restriction on which is that they must not pair with the other chromosomes of the wheat complement. These are a modified 4A–2R translocation chromosome, part of the cereal rye genome, chromosome 4 of barley, and chromosome 4 of diploid wheat. It appears that chromosome 4A originated elsewhere than from diploid wheat; thus the A genome of common wheat arose from at least two species. As detected by M. A. Hossain and the author, part of the genetic material for male fertility on 4A has a counterpart on rye chromosome 2R. An exchange between the chromosomes of homoeologous groups 2 and 4 appears to have occurred, perhaps at the diploid level.

The manipulation of chromosomes or chromosome segments is assuming an increasingly important role in plant breeding. These manipulations, which have been used rather extensively in the improvement of common wheat (*Triticum aestivum* L.), include uncontrolled events induced by mutagenic agents and events that are under genetic control. The latter may be divided into two categories, depending on whether they are based upon the unusual genetic characteristics of chromosomes 5B or 4A. The impact of these three categories of manipulation on plant improvement is considered, as they collectively illustrate the developments in this general area of research.

Uncontrolled events induced by mutagenic agents

Considerable emphasis has been placed on the introduction of 'alien' genetic material, i.e. material from other than the host species, in the improvement of hexaploid wheat. Wide crossing has mostly involved species from the subtribe Triticinae Melderis; however, successful crosses have been made beyond this boundary, e.g. those involving wheat and various species of *Hordeum* L., including *H. vulgare* L. (Islam *et al.* 1975).

The resultant amphiploids, with one exception, have not been directly used in agriculture. The exception, triticale (× *Triticosecale* Wittmack), resulting from crosses of wheat and rye (*Secale* L.), is being used in agriculture on an increasing scale.

Single-chromosome addition lines and substitution lines have not played a significant role in agriculture in that the alien genetic material added or substituted is too great and causes problems in production characters. The notable exception to this is the series of varieties that include the addition of rye (*Secale cereale* L.) chromosome 1R or the substitution of it for chromosome 1B (Mettin *et al.* 1973; Zeller 1973).

Transfer of less than a full chromosome to the recipient species has been accomplished by the use of ionizing radiation. The first transfer of this type involved transfer of a segment of an *Aegilops umbellulata* Zhuk. chromosome, bearing resistance to wheat leaf rust (*Puccinia recondita* Rob. ex Desm.), to chromosome 6B of wheat (Sears 1956).

The most successful transfer of this type involves the translocation of a segment of an *Agropyron elongatum* (Host) Beauv. chromosome that carries resistance to wheat stem rust (*Puccinia graminis tritici* Erikss.) to chromosome 6A of wheat (Knott 1961). This resistance has been incorporated into Australian wheat varieties Eagle (Martin 1971), Kite (Fisher & Martin 1974), Jabiru (Fisher & Syme 1977) and Avocet (R. H. Martin, personal communication), which in recent years have collectively constituted approximately 700 kha or 7 % of the Australian wheat crop.

However, most of these translocations (see Driscoll (1975a, 1976a) for listings) have not been successfully incorporated into widely grown varieties of wheat. The reasons for this are loss of important wheat genetic material and/or gain of deleterious alien genetic material in the exchange. For example, the Transec translocation involves transfer of a segment of rye chromosome 2R, bearing resistance to wheat leaf rust and to wheat powdery mildew (*Erysiphe graminis tritici* March.), to chromosome 4A of wheat (Driscoll & Jensen 1964). This translocation has not been incorporated into a released variety of wheat despite the fact that it has been available to plant breeders for over 15 years. The translocation is usually associated with lowered yield. It has been proposed by Shepherd (1977) that if selection is imposed on this type of material under specific conditions it may be possible to break this association. This proposal includes selection of large numbers of F_3, or equivalent, plants that are homozygous for the alien translocation and subjecting derived F_5, or equivalent, progenies to yield tests. The rationale is that many pairs of alleles will assort between F_3 and F_5, and considering that all lines are homozygous for the alien segment, selection for high yield will involve selection for a genetic background that accommodates the alien segment. K. W. Shepherd (personal communication) is examining this concept with spontaneously arising transfers of segments of rye chromosome 1R carrying wheat stem rust resistance and I am attempting to select an accommodating genetic background for the Transec translocation by this method. In the latter study 950 F_5 (or F_7) selections are being subjected to replicated yield testing in the 1980 season.

If these experiments prove to be successful they will be of some significance, as this may define the special treatment that can or must be applied to alien transfers to ensure their use in agriculture. It will be particularly important if high-yielding types are found that on further crossing provide high-yielding segregants in reasonably high frequencies. Special handling of that particular translocation may then not be necessary in further crosses. This would be more likely to arise if the improvement were due to an infrequent change in the alien segment itself or a simply inherited, but major, change in the genetic background, such as on a homoeologue of the translocation chromosome.

Alien transfers of the type being discussed may also come into greater prominence if hybrid wheat becomes significant. If an alien segment is present in only one parent, it is present in single rather than in double dose in the hybrid, and the replaced wheat segment is reintroduced (in single dose) by the other parent.

Despite these reasons why translocations induced by ionizing radiation may be more successfully used in the future, production of further transfers will be predominantly, if not entirely, engineered by way of *ph* mutants.

Manipulations involving chromosome 5B

The important role of chromosome 5B in control of recombination is discussed elsewhere (Riley *et al.*, this symposium); thus only one aspect will be considered here, namely the manipulations associated with the use of the *ph* mutants isolated by Riley (1968) and Sears (1977). It is expected that use of a *ph* mutant will result in more precise exchanges in terms of being homoeologous exchanges.

The extent of homoeologous pairing of these mutants has been examined mathematically (Driscoll *et al.* 1979, 1980; Driscoll 1979) and in one case (Sears 1977) the degree of pairing in an intergeneric hybrid exceeds that of the same hybrid involving nullisomic 5B.

Sears (1980) has outlined a method in which the exchange is also more precise in the length and positioning of the alien segment in terms of being short and intercalary. The method involves crossing a monosomic 5B, ditelosomic alien substitution line with a homozygous *ph* line. Offspring with 41 chromosomes involve homoeologous pairing, and the alien telochromosome and its wheat homoeologue that it had replaced are devoid of homologues and are therefore available for homoeologous exchange. There are two types of exchange chromosomes that bear the gene being transferred: an alien telocentric chromosome with a wheat segment distal to the gene, and a two-armed wheat chromosome bearing a terminal alien segment that carries the gene. These two types are specifically identified, after crossing, by the appropriate wheat ditelocentric stock, namely that involving the wheat arm involved in the exchanges. Furthermore, the extent of telocentric pairing in this generation serves as a guide to the length of translocated segment. Intercrossing lines that carry the two types of exchange brings together two chromosomes with a short alien segment in common (see figure 1). Exchange in this segment produces a chromosome with an intercalated alien segment bearing the gene being transferred.

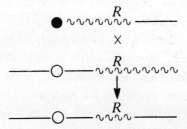

Figure 1. Crossing over between the two types of exchange chromosomes to produce the intercalated chromosome. *R* is the gene being transferred (after Sears 1980).

Manipulations involving chromosome 4A

Chromosome 4A of *Triticum aestivum* and *Triticum durum* Desf. has a number of interesting features. Sears (1966) observed that all three nullisomics of homoeologous group 4 in hexaploid wheat are male sterile and generally non-vigorous. Further, nullisomic 4A tetrasomic 4B (or 4D) is quite normal in all respects except that it is male sterile, whereas the reciprocal, nullisomic 4B (or 4D) tetrasomic 4A, is normal in all respects including male fertility. Hence 4A has the gene(s) for male fertility that both 4B and 4D have and in addition it has a gene(s) for male fertility that neither 4B nor 4D have.

Secondly, 4A of hexaploid wheat is unusual in its pairing relationship with diploid *Triticum* species. Chapman *et al.* (1976) observed no pairing of the telocentric chromosomes in hybrids of double ditelocentric 4A of hexaploid wheat × *T. urartu* Thum. Telocentrics of the other six A-genome chromosomes of hexaploid wheat paired with chromosomes of *T. urartu*. Similar observations were made by Dvořák (1976), who employed ditelocentric 4Aα. Earlier, Chapman & Riley (1966) reported 48% pairing of telocentric 4Aα in crosses to *T. thaoudar* Reut. However, K. W. Shepherd (personal communication) observed no pairing of telocentrics 4Aα and 4Aβ in crosses with *T. thaoudar*, *T. boeticum* Boiss., *T. aegilopoides* Bal. and *T. monococcum* L. Further, M. D. Hossain (personal communication) observed no pairing of the entire 4A of hexaploid wheat in crosses to *T. monococcum*. In this case, 4A of hexaploid wheat was recognized by means of N-banding (Gerlach 1977; Jewell 1979). From the above observations it can be concluded that 4A of hexaploid wheat originated elsewhere than from diploid wheat or it has been modified so dramatically it no longer pairs with, or is seen paired at metaphase I with, a chromosome of diploid wheat (presumably this can be referred to as chromosome 4 of diploid wheat as the other six chromosomes pair with the other six chromosomes of the A genome of hexaploid wheat).

Thirdly, the N-banding pattern of chromosome 4A makes it conspicuously different from the other six chromosomes of the A genome. Gerlach (1977) demonstrated that in hexaploid wheat all seven chromosomes of the B genome are heavily and distinctly banded, chromosome 7A has faint banding and chromosome 4A is heavily banded. Chromosome 4A has heavy banding on each side of the centromere and a subterminal band in the β arm. The same pattern of 4A is seen with polypyrimidine–polypurine satellite *in situ* hybridization (Dennis *et al.* 1980). Chromosome 4 of diploid wheat is not banded (Gerlach 1977). This evidence, in conjunction with that presented earlier, leads one to conclude that it is most likely that chromosome 4A of hexaploid wheat was derived elsewhere than from diploid wheat.

A number of manipulations of chromosome 4A have been carried out with interest in male sterility for plant breeding purposes. The male sterility of nullisomic 4A is due to loss of a gene or genes on the α arm, since ditelocentric 4Aα is fertile and ditelocentric 4Aβ is male sterile. There are five chromosomes of hexaploid wheat, namely 4A, 4B, 5A, 5B and 5D, of which the nullisomic is male sterile, one telocentric is male sterile and the other telocentric is male fertile (E. R. Sears, personal communication). These five were chosen for an induced-mutation study to isolate a recessive mutation for male sterility.

Euploid pollen was subjected to 250 or 500 R† of γ-rays or X-rays, at about the time of pollen shedding, and applied to emasculated monosomics 4A, 4B, 5A, 5B or 5D (Driscoll & Barlow 1976). Approximately three-quarters of the F_1s will not receive the monosome from the

† $1 R = 2.58 \times 10^{-4}$ C kg^{-1}.

female parent, and if in these cases a recessive mutation is induced in the corresponding chromosome of the pollen, the recessive mutation will be expressed in the F_1. Of 795 F_1 plants (62 of which involved monosomic 4A as female parent) produced in this way, one male-sterility mutant was isolated, confirmed and maintained (Driscoll 1976b). This mutant has been registered as 'Cornerstone' (Driscoll 1977a) and has been comparatively analysed with two other recessive male-sterility mutants, referred to as 'Pugsley's' male sterile (Pugsley & Oram 1959) and the 'Probus' male sterile (Fossati & Ingold 1970). The three recessive mutants are allelic and are located on chromosome arm 4Aα (Driscoll 1975b; Barlow & Driscoll 1980). The Probus and the Cornerstone mutants are located at least 50 map units from the centromere. Each of these two mutant chromosomes when paired with the telocentric 4Aα undergo considerable desynapsis (Barlow & Driscoll 1980).

One other aspect of these studies is important. Progenies resulting from crossing monosomic 4B with the Probus (or Cornerstone) heterozygote contain individuals with an intermediate phenotype, in which many anthers fail to dehisce but others liberate functional pollen. These plants are regarded as being monosomic for chromosome 4B and heterozygous for the 4A mutant. The fact that the intermediate type was observed with the two mutants induced with ionizing radiation (Probus and Cornerstone) and not with the spontaneous mutant (Pugsley's) is interpreted as indicating that 4Aα bears two genes for male fertility, one of which is duplicated on 4B, and that both genes are deleted in the Probus and Cornerstone mutants but only the duplicated 4A gene is changed in the case of the spontaneous Pugsley's mutant. This is consistent with the observations on the nullisomic–tetrasomics of homoeologous group 4 (Sears 1966) mentioned above.

It is significant that the first three male-sterility mutants studied in hexaploid wheat fail to complement one another. The unduplicated 4A gene for male fertility is more likely to give rise to detectable mutations. However, it should be noted that a large number of male-sterility mutants have been induced with ethyl methanesulphonate by Franckowiak et al. (1976). These involved four non-allelic recessive mutants (Sasakuma et al. 1978). These mutants may be modified alleles that actively compete with their homoeoalleles. A similar situation has been experienced with chlorophyll mutants in hexaploid wheat (Sears & Sears 1968; Pettigrew & Driscoll 1970). In this case, almost no chlorophyll mutants arise with ionizing irradiation, which may predominantly induce deletions; however, chemical mutagens induce mutants, such as chlorina mutants, that involve intragenic changes.

The Cornerstone mutant is being used in composite crosses to render the wheat plant open-pollinating for a series of generations (Driscoll 1977b, 1980).

What may prove to be of greater significance are the attempts that are being made to use Cornerstone in production of hybrid wheat. Hybrid wheat production by using cytoplasmic male sterility, specifically *Triticum timopheevi* Zhuk., as described by Schmidt et al. (1962), has essentially been discontinued because of difficulties in fertility restoration. An alternative method of producing hybrid wheat, described by Driscoll (1972), involves a chromosomal male-sterility mutant and an additional chromosome that bears a gene(s) for fertility restoration and, secondly, does not pair at meiosis with the other chromosomes.

Cornerstone is suitable for this system; however, isolation of the additional chromosome with the two characteristics mentioned above is proving to be difficult. Similar to the nullisomic–tetrasomic situation with chromosome 4A, as mentioned above, substitution of an alien chromosome for chromosome 4A can give rise to a vigorous but male-sterile line. For example, substitution of chromosome 4E of diploid *Agropyron elongatum* for 4A results in this type of individual (Dvořák 1980). Thus it is assumed that an alien chromosome of this type will not restore fertility to the Cornerstone mutant, which is regarded as being a terminal deletion of chromosome arm 4Aα (Barlow & Driscoll 1980). There is one *Aegilops* chromosome that substitutes successfully for chromosome 4A in that it results in a male-fertile substitution line. This is chromosome $4S^1$ of *A. longissima* S. & M. as demonstrated by Maan (1975, 1976, 1980); however, this chromosome has a most peculiar characteristic: when present monosomically, male and female gametes not bearing the *Ae. longissima* chromosome are precluded from functioning. This behaviour has also been observed with a chromosome of *Ae. sharonensis* Eig by Maan (1975) and Chapman & Miller (1978); however, this may be the same chromosome, namely $4S^1$, as both *Ae. longissima* and *Ae. sharonensis* are regarded as having the same genome, i.e. S^1.

This gametocidal behaviour is not, however, unique to group 4, since a group 3 chromosome of *Ae. triuncialis* L. var. *triuncialis* (genomes CCC^uC^u), of *Ae. caudata* L. (CC) and of a synthetic *Ae. triuncialis* obtained by crossing *Ae. caudata* × *Ae. umbellulata* (C^uC^u) has a similar property (Endo & Tsunewaki 1974; Endo & Katayama 1978; Endo 1978). This chromosome, which is most probably chromosome 3C of *Ae. caudata* and substitutes well for 3A, 3B and 3D, but unsuccessfully for 4A, is different from the previous case in that its gametocidal action is partial rather than absolute (Endo 1978).

Endo (1980) also reports a gametocidal action of a chromosome of *Ae. cylindrica* Host (CCDD). It is also reported that although this chromosome is probably a C-genome chromosome it may not be the same chromosome as the one described in *Ae. triuncialis*.

As (partial) gametocidal action of chromosomes other than group 4 chromosomes have been recognized, the gametocidal action and the fertility restoration of $4S^1$ probably result from two distinct genetic activities. Therefore the two activities may result from two genes (or two groups of genes) that are separable. Isolation of a $4S^1$ chromosome that restores the fertility but does not have the gametocidal activity would provide an alien chromosome useful for the production of hybrid wheat by the method mentioned above. Isolation of a $4S^1$ chromosome with these properties could be attempted by an appropriately designed irradiation experiment; alternatively, it may occur naturally, and perhaps further accessions of *Ae. longissima* and *Ae. sharonensis* should be screened for this purpose.

If a fertility-restoring, non-gametocidal $4S^1$ chromosome (I shall refer to it as chromosome β) is isolated, the isolation of the reciprocal type, i.e. a gametocidal, non-restoring $4S^1$ chromosome (chromosome α) would allow the examination of an alternative method of producing hybrid wheat, which even though no abnormal cytoplasm is involved, is quite similar to the cytoplasmic system of producing hybrids. A further requirement would be the absence of crossing-over between the fertility-restoring and the gametocidal genes. A line with a pair of α chromosomes would be sterile and equivalent to an A line of the cytoplasmic system (Jones & Davis 1944). A line with one α and one β chromosome would be fertile and equivalent to a B line. The αα and αβ lines could be grown in mixture and only α-bearing male and female gametes would function and the mixture could be harvested in bulk to give only αα types. This would have advantages in pollen distribution. Maintenance of the αβ line, incidentally, would require pollination of an αα line with a ββ line, the latter being a self-reproducing type. The unusual character about the αα type, which could be produced in bulk, is that it carries two doses of a gametocidal chromosome. Hence the male parent of the hybrid production would also have to be αα type or else the hybrid crop will be an α type with the majority of eggs non-functional. Hence the hybrid crop would have to include a pair of α chromosomes, and this may negate any potential hybrid vigour. Thus, perhaps, isolation of an α chromosome and use in the XYZ system (Driscoll 1972) would be preferable.

Four other possibilities are being examined for use in the XYZ system. The first is the Transec chromosome. It involves a segment of rye chromosome 2R translocated to the β arm of 4A and located one map unit from the centromere (Driscoll & Bielig 1968). In crosses to Cornerstone the two 4A chromosomes are paired at metaphase I of the F_1 in only 3% of pollen mother cells (Barlow & Driscoll 1980). A line was produced that possessed two doses of Cornerstone 4A and one dose of Transec 4A, and this line involved trivalent formation at metaphase I in only 1% of pollen mother cells. This seemed low enough for this to be a successful Y line. However, from a linkage study between the Cornerstone and Transec 4A chromosomes, considerable crossing-over occurs between these chromosomes. In fact the rye segment on the β arm and the male-sterility deletion map 20 crossover units apart (Barlow & Driscoll 1980). Obviously, considerable desynapsis as well as asynapsis is involved in this case and these have been estimated at 37% and 60% respectively, from a formula derived by Driscoll (1978b). The desynapsis of this bivalent is due to the abnormality of both pairs of ends of this bivalent: the α end is heterozygous for the Cornerstone mutant, which is presumably a terminal deletion, and the β end is heterozygous for 2R and 4A segments. Considering that the 2R segment maps one crossover unit from the centromere, the 20% crossing-over in this bivalent presumably almost all occurs in the α arm.

Attempts are being made with the use of X-rays to reduce this pairing in the α-arm. The Transec chromosome has been irradiated (250 R at the time of pollen shed) and applied to ditelocentric 4Aα. The irradiated Transec 4A chromosomes are being scored for their abilities to remain paired until metaphase I with telocentric 4Aα, rather than with the Cornerstone 4A, because the non-irradiated Transec 4A seldom remains paired with the Cornerstone 4A and any reduction in zygotene pairing would not be detected in such a test. Of 36 irradiated Transec 4A chromosomes, 34 were observed at metaphase I to be paired with telocentric 4Aα with normal frequency (i.e. 78% paired) and 2 paired less frequently, namely 6/23 (26%) and 6/18 (33%). The last two have been selfed to produce lines that possess a pair of irradiated-Transec 4A chromosomes, to test for retention of the male fertility genes. These lines are also being crossed to Cornerstone to determine if their modified chromosome pairs negligibly with Cornerstone 4A. Of course this will have to be determined by a linkage study of the wheat leaf rust resistance gene carried by the 2R segment and the male-sterility allele, since the non-irradiated Transec and Cornerstone 4A chromosomes are seen paired at metaphase I only rarely and observations on the earlier stages of meiosis are too difficult.

The second source of fertility restoration that is being examined is cereal rye (*S. cereale*). The Cornerstone mutant has been transferred to the tetraploid level by backcrossing to *T. durum* and the hexaploid and tetraploid sterile stocks incorporated into octoploid amphiploids and hexaploid amphiploids, respectively, with rye. Both amphiploids have normal anthers and liberate normal pollen (Hossain & Driscoll 1980). Thus the entire rye genome is capable of restoring male fertility. These amphiploids do not have full seed set; however, many amphiploids of these types that do not involve a male-sterility mutation do not have full seed set. The heptaploid containing a single dose of the rye genome added to the hexaploid, homozygous for the male-sterility mutation, is also male fertile; thus the rye genome in single dose is capable of restoring fertility. It is not surprising that this is so; presumably all higher plants, or at least those of a restricted taxonomic group, construct functional anthers and pollen grains in similar, if not identical, ways. The differences could involve the packaging of the genes, as to whether a given pair or group of genes are linked in one species but not so in another species. The packaging in rye with respect to the genes on 4Aα, that have been deleted in Cornerstone, is being analysed by backcrossing the heptaploid to hexaploid Cornerstone and analysing the BC_2 progeny for rye-chromosome content and for male fertility or sterility. The identity of the rye chromosomes is being ascertained by C-banding (Gill & Kimber 1974) and the electrophoretic analysis of esterases and alcohol dehydrogenases, which are used as markers for chromosomes 3R and 4R, respectively (Barber et al. 1968; Tang & Hart 1975).

Although analysis of this BC_2 family is incomplete, a number of findings have already emerged (M. D. Hossain & C. J. Driscoll, unpublished). The possibility that 5R may carry the restorer gene(s) for the Cornerstone 4A mutation had its basis in the observation that the spontaneous wheat line W70a86 (Blaukorn) is male fertile and involves substitution of rye chromosome 5R for wheat chromosome 4A (Zeller & Baier 1973). However, isolation of a BC_2 plant with 5R added to Cornerstone is male sterile; thus W70a86 (Blaukorn) may be male fertile for a reason other than the presence of 5R, or the Blaukorn 5R is a distinctive one.

On the positive side, it is significant that rye chromosome 2R added to Cornerstone results in partial fertility (51 % seed set in primary and secondary florets). It was proposed earlier that Cornerstone involves the deletion of two genes for male fertility, only one of which is duplicated on the 4B and 4D homoeologues. Perhaps a gene comparable with the unduplicated gene on 4A is present on rye chromosome 2R. This chromosome is also related to chromosomes 2A, 2B and 2D of hexaploid wheat (Sears 1968). Interchange of chromosomes 2A, 4A and 6B was postulated by Sears (1966) on the basis of incomplete compensation in particular nullisomic–tetrasomic lines. This strongly suggests that 4A of hexaploid (and tetraploid) wheat arose from a chromosome that also has a relation with homoeologous group 2 of hexaploid wheat. This, and the fact that chromosome 4A of hexaploid wheat N-bands in a manner similar to a B-genome chromosome, suggests that the A genome of hexaploid wheat has a complex origin that includes diploid *Triticum* and at least one other genus, with chromosomes that heavily N-band.

Perhaps a second rye chromosome that will partly restore fertility will occur in the BC_2 population referred to above. This seems particularly likely when one considers that Koller & Zeller (1976) reported 4 and 19 % fertility restoration in monosomic 4R substitution for 4A and ditelosomic 4RS substitution for 4A, respectively. If two rye chromosomes are involved in restoring fertility it may be possible to combine their partial restoring activities into one chromosome by centric fusion (Sears 1972). The homoeology of a segment of 2R and a segment of 4A allows the possibility that the exchange between these two chromosomes in the Transec chromosome is a segmental homoeologous one.

It would, of course, be easier to find a chromosome 'off the peg'. One possibility of this is the third chromosome that is being examined in this context, namely chromosome 4 of barley (*H. vulgare*). Crosses of common wheat with cultivated barley and production of six of the possible seven addition lines have been achieved by Islam et al. (1981). Using N-banding of Gerlach (1977), Islam (1980) identified his addition line A as involving barley chromosome 4, as numbered in the trisomic series by Tsuchiya et al. (1960). Barley chromosome 4 was found to produce an alcohol dehydrogenase that is electrophoretically different to those produced by the group 4 chromosomes of wheat (Hart et al. 1980).

An addition line with a pair of barley chromosome 4, including one chromosome with a markedly different arm ratio, was isolated and crossed to monosomic 4A. In a subsequent generation, monitoring of alcohol dehydrogenases led to the isolation of a disomic substitution line in which 4A is replaced by a pair of barley 4 with the modified arm ratio. This disomic substitution line is male fertile (A. K. M. R. Islam and K. W. Shepherd, personal communication). Work is under way to determine whether a normal barley 4 will also produce a male-fertile 4A substitution line. Thus barley chromosome 4, or perhaps more specifically the structurally modified barley chromosome 4, may prove to be a useful chromosome for the XYZ system. Development of X and Y lines are now in progress.

The fourth chromosome that is being examined for fertility restoration came into consideration because of a happy accident. In attempting to transfer the Cornerstone mutant for male sterility from tetraploid wheat to diploid wheat, male-sterile tetraploid was pollinated with *Triticum monococcum*. The F_1 was treated with colchicine in order to increase the mitotic index in root tips; however, it also had the additional, unplanned effect of producing a doubled sector consisting of two entire spikes. These contained genomes AAA**A**BB, where the bold **A** is from *T. monococcum*. This doubled sector was fully fertile (M. D. Hossian & C. J. Driscoll, unpublished). The point of interest is that in the undoubled F_1 (i.e. A**A**B) only six of the seven **A**-genome chromosomes pair with A-genome homologues, N-banding these meiotic figures reveals that chromosome 4**A** of the **A** genome is unpaired. Thus chromosome 4 of the **A** genome does not pair with the chromosomes of hexaploid wheat. If full fertility is restored by genes on chromosome 4 of the **A** genome, it will be a useful chromosome for this system. *T. monococcum* is being crossed to hexaploid Cornerstone with the purpose of producing a chromosome 4 addition line on a male-sterile background.

In time, a suitable fertility-restoring chromosome will be isolated, and it may well come from one of the four mentioned above. This research is being persevered with because of two special features of producing hybrids in this way. First, the normal cytoplasm would be involved in the hybrid crop and, secondly, no alteration, in terms of sterility–fertility, is made with the male parent of the hybrid. Once X, Y and Z lines are produced in a given genotype, that genotype can be readily tested with a large number of other genotypes to search for combinations that have sufficient heterosis for yield.

I have discussed chromosome manipulations in wheat and a number of its relatives with the aim of introducing alien genes to wheat or to produce hybrid wheat. Many other examples with different species and different objectives could have been assembled. The examples chosen do, however, illustrate one very important concept, and that is that advanced manipulation of chromosomes depends heavily on a basic knowledge of the relations between chromosomes. This knowledge is well advanced with wheat, which reflects the fact that many aneuploids are available within this species. The examples given reflect in a general way the extent of development of the manipulation of chromosomes of higher plants for purposes of plant breeding.

Discussions with K. W. Shepherd and E. R. Sears are gratefully acknowledged.

References (Driscoll)

Barber, H. N., Driscoll, C. J., Long, P. M. & Vickery, R. S. 1968 Protein genetics of wheat and homoeologous relationships of chromosomes. *Nature, Lond.* **218**, 450–452.

Barlow, K. K. & Driscoll, C. J. 1980 Linkage studies involving two chromosomal male-sterility mutants in hexaploid wheat. (In the press.)

Chapman, V. & Miller, T. E. 1978 Alien chromosome addition and substitution lines. *Plant Breeding Institute, Cambridge, Annual Report*, pp. 124–125.

Chapman, V. & Riley, R. 1966 The allocation of the chromosomes of *Triticum aestivum* to the A and B genomes and evidence of genome structure. *Can. J. Genet. Cytol.* **8**, 257–263.

Chapman, V., Miller, T. E. & Riley, R. 1976 Equivalence of the A genome of bread wheat with that of *Triticum urartu*. *Genet. Res. Camb.* **27**, 69–76.

Dennis, E. S., Gerlach, W. L. & Peacock, W. J. 1980 Identical polypyrimidine–polypurine satellite DNAs in wheat and barley. *Heredity, London.* **44**, 349–366.

Driscoll, C. J. 1972 XYZ system of producing hybrid wheat. *Crop. Sci.* **12**, 516–517.

Driscoll, C. J. 1975a First compendium of wheat–alien chromosome lines. *A. Wheat Newsl.* **21**, 16–32.

Driscoll, C. J. 1975b Cytogenetic analysis of two chromosomal male-sterility mutants in hexaploid wheat. *Aust. J. biol. Sci.* **28**, 413–416.

Driscoll, C. J. 1976a Second compendium of wheat–alien chromosome lines. *A. Wheat Newsl.* **22**, 4–5.

Driscoll, C. J. 1976b Induction and use of chromosomal male sterility in common wheat. In *Experimental mutagenesis in plants* (ed. K. Filev), pp. 186–188. Varna.

Driscoll, C. J. 1977a Registration of Cornerstone male-sterile wheat germplasm. *Crop Sci.* **17**, 190.

Driscoll, C. J. 1977b Reorganizing existing variability in wheat and related species. In *3rd International Congress of Society for Advancement of Breeding Researches in Asia and Oceania/Australian Plant Breeding Conference*, Canberra, vol. 1, pp. 2(a)1–2(a)8.

Driscoll, C. J. 1978 Mapping alien segments. In *Cytogenetics and crop improvement* (ed. R. B. Singh). Varanasi. (In the press.)

Driscoll, C. J. 1979 Mathematical comparison of homologous and homoeologous chromosome configurations and the mode of action of the genes regulating pairing in wheat. *Genetics, Princeton*, **92**, 947–951.

Driscoll, C. J. 1980 New approaches to wheat breeding. In *Wheat Science: today and tomorrow* (ed. W. J. Peacock & L. T. Evans). Canberra. (In the press.)

Driscoll, C. J. and Barlow, K. K. 1976 Male sterility in plants: induction, isolation and utilization. In *Induced mutations in cross-breeding*, pp. 123–131. Vienna: I.A.E.A.

Driscoll, C. J. & Bielig, L. M. 1968 Mapping of the Transec wheat-rye translocation. *Can. J. Genet. Cytol.* **10**, 421–425.

Driscoll, C. J., Bielig, L. M. & Darvey, N. L. 1979 An analysis of frequencies of chromosome configurations in wheat and wheat hybrids. *Genetics, Princeton* **91**, 755–767.

Driscoll, C. J. & Jensen, N. F. 1962 Characteristics of leaf rust resistance transferred from rye to wheat. *Crop Sci.* **4**, 372–374.

Driscoll, C. J., Gordon, G. H. & Kimber, G. 1980 Mathematics of chromosome pairing. *Genetics, Princeton* **95**, 159–169.

Dvořák, J. 1976 The relationship between the genome of *Triticum urartu* and the A and B genomes of *Triticum aestivum*. *Can. J. Genet. Cytol.* **18**, 371–377.

Dvořák, J. 1980 Homoeology between *Agropyron elongatum* chromosomes and *Triticum aestivum* chromosomes. *Can. J. Genet. Cytol.* **22**, 237–259.

Endo, T. C. 1978 On the *Aegilops* chromosomes having gametocidal action on common wheat. In *5th International Wheat Genetics Symposium*, vol. 1, pp. 306–314.

Endo, T. R. 1980 Selective gametocidal action of a chromosome of *Aegilops cylindrica* in a cultivar of common wheat. *Wheat Inform. Serv., Kyoto* nos. 27 and 28, pp. 32–35.

Endo, T. R. & Tsunewaki, K. 1975 Sterility of common wheat with *Aegilops triuncialis* cytoplasm. *J. Hered.* **66**, 13–18.

Fisher, J. A. & Martin, R. H. 1974 Kite, Condor and Egret – three new wheats. *Agr. Gazette N.S.W.* **85** (2), 10–13.

Fisher, J. A. & Syme, J. R. 1977 Register of cereal and *Linum* varieties in Australia. Wheat Jabiru. *J. Aust. Inst. agric. Sci.* no. 173.

Fossati, A. & Ingold, M. 1970 A male sterile mutant in *Triticum aestivum*. *Wheat Inform. Serv. Kyoto* **30**, 8–10.

Franckowiak, J. D., Maan, S. S. & Williams, N. D. 1976 A proposal for hybrid wheat utilizing *Aegilops squarrosa* L. cytoplasm. *Crop. Sci.* **16**, 725–728.

Gerlach, W. L. 1977 N-banded karyotypes of wheat species. *Chromosoma* **62**, 49–56.

Gill, B. S. & Kimber, G. 1974 The giemsa C-banded karyotype of rye. *Proc. natn. Acad. Sci. U.S.A.* **71**, 1247–1249.

Hart, G. E., Islam, A. K. M. R. & Shepherd, K. W. 1980 Use of isozymes as chromosome markers in the isolation and characterization of wheat–barley chromosome addition lines. *Genet. Res., Camb.* (In the press.)

Hossain, M. D. & Driscoll, C. J. 1980 Transfer of Cornerstone male-sterility mutant to tetraploid wheat and hexaploid and octoploid triticales. *Can. J. Genet. Cytol.* (In the press.)

Islam, A. K. M. R. 1980 Identification of wheat-barley addition lines with N-banding of chromosomes. *Chromosoma* **76**, 365–373.

Islam, A. K. M. R., Shepherd, K. W. & Sparrow, D. H. B. 1975 Addition of individual barley chromosomes to wheat. In *3rd International Barley Genetics Symposium*, Garching, G.D.R., pp. 260–270.

Islam, A. K. M. R., Shepherd, K. W. & Sparrow, D. H. B. 1981 Isolation and characterization of euplasmic wheat-barley chromosome addition lines. *Heredity, Lond.* (In the press.)

Jewell, D. C. 1979 Chromosome banding in *Triticum aestivum* cv. Chinese Spring and *Aegilops variabilis*. *Chromosoma* **71**, 129–134.

Jones, H. A. & Davis, G. N. 1944 Inbreeding and heterosis and their relation to the development of new varieties of onions. *U.S. Dept. Agric. tech. Bull.* no. 874, pp. 1–28.

Knott, D. R. 1961 The inheritance of rust resistance. VI. The transfer of stem rust resistance from *Agropyron elongatum* to common wheat. *Can. J. Pl. Sci.* **41**, 109–123.

Koller, O. L. & Zeller, F. J. 1976 The homoeologous relationships of rye chromosomes 4R and 7R with wheat chromosomes. *Genet. Res., Camb.* **28**, 177–188.

Maan, S. A. 1975 Exclusive preferential transmission of an alien chromosome in common wheat. *Crop Sci.* **15**, 287–292.

Maan, S. S. 1976 Alien chromosome controlling sporophytic sterility in common wheat. *Crop Sci.* **16**, 580–583.

Maan, S. A. 1980 Alteration of sporophytic sterility mechanism in wheat. *J. Hered.* **71**, 75–82.

Martin, R. H. 1971 Eagle – a new wheat cultivar. *Agr. Gazette N.S.W.* **82**, 207.

Mettin, D., Blüthner, W. D. & Schlegel, G. 1973 Additional evidence on spontaneous 1B/1R wheat-rye substitutions and translocations. In *4th International Wheat Genetics Symposium*, Columbia, Missouri, pp. 179–184.

Pettigrew, R. & Driscoll, C. J. 1970 Cytogenetic studies of a chlorophyll mutant of hexaploid wheat. *Heredity, Lond.* **25**, 650–655.

Pugsley, A. T. & Oram, R. N. 1959 Genic male sterility in wheat. *Aust. Pl. Breed. Genet. Newsl.* **14**, 10–11.

Riley, R. 1968 The basic and applied genetics of chromosome pairing. In *3rd International Wheat Genetics Symposium*, Canberra, pp. 185–195.

Sasakuma, T., Maan, S. S. & Williams, N. D. 1978 EMS-induced male-sterile mutants in euplasmic and alloplasmic common wheat. *Crop Sci.* **18**, 850–853.

Schmidt, J. W., Johnson, V. A. & Maan, S. S. 1962 Hybrid wheat. *Nebraska exp. stn. Q.* **9** (3), 9.

Sears, E. R. 1956 The transfer of leaf-rust resistance from *Aegilops umbellulata* to wheat. *Brookhaven Symp. Biol.* **9**, 1–22.

Sears, E. R. 1966 Nullisomic-tetrasomic combinations in hexaploid wheat. In *Chromosome manipulations and plant genetics* (ed. R. Riley & K. R. Lewis), pp. 29–45. Edinburgh: Oliver & Boyd.

Sears, E. R. 1968 Relationships of chromosomes 2A, 2B and 2D with their rye homoeologue. In *3rd International Wheat Genetics Symposium*, Columbia, Missouri, pp. 53–61.

Sears, E. R. 1972 Chromosome engineering in wheat. *Stadler Symp.* **4**, 23–38.

Sears, E. R. 1977 An induced mutant with homoeologous pairing in common wheat. *Can. J. Genet. Cytol.* **19**, 585–593.

Sears, E. R. 1980 Transfer of alien genetic material to wheat. In *Wheat science: today and tomorrow* (ed. W. J. Peacock & L. T. Evans). Canberra. (In the press.)

Sears, L. M. S. & Sears, E. R. 1968 The mutants *Chlorina*-1 and Hermsen's Virescent. In *3rd International Wheat Genetics Symposium*, Columbia, Missouri, pp. 299–304.

Shepherd, K. W. 1977 Utilization of a rye chromosome arm in wheat breeding. In *3rd International Congress of Society for Advancement of Breeding Researches in Asia and Oceania/Australian Plant Breeding Conference*. Canberra, vol. 1, pp. 2(a) 16–2(a) 20.

Tang, K. W. & Hart, G. E. 1975 Use of isozymes as chromosome markers in wheat-rye addition lines and in triticale. *Genet. Res., Camb.* **26**, 187–201.

Tsuchiya, T., Hayashi, J. & Takahashi, R. 1960 Genetic studies in trisomic barley. II. Further studies on the relationship between trisomics and the genetic linkage groups. *Jap. J. Genet.* **35**, 153–160.

Zeller, F. J. 1973 1B/1R wheat-rye chromosome substitutions and translocations. In *4th International Wheat Genetics Symposium*, Columbia, Missouri, pp. 209–221.

Zeller, F. J. & Baier, A. C. 1973 Substitution des Weizenchromosomenpaares 4A durch das Roggenchomosomenpaar 5R in dem Weihenstephaner Weizenstamm W70a86 (Blaukorn). *Z. PflZücht.* **70**, 1–10.

Discussion

T. E. MILLER (*Plant Breeding Institute, Cambridge, U.K.*). Professor Driscoll has shown that chromosome 4A in hexaploid wheat may not be an A genome chromosome from the diploid wheats, and recent evidence at the Plant Breeding Institute would support this. Does Professor

Driscoll consider that we should therefore question the integrity of genomes and hence their evolutionary origin in the polyploid cereals, particularly the B genome of wheat?

C. J. DRISCOLL. From the data discussed, it appears that the origin of the A genome of hexaploid wheat involved more than one diploid species. Diploid wheat apparently contributed five or, more likely, six chromosomes; however, chromosome 4A seems to have been derived from another species. Thus both the A genome and the B genome of hexaploid wheat appear to have had a complex origin.

Postscript. Prospects for hybrid wheat may be greater than is commonly envisaged. Most studies on heterosis in wheat have involved the *T. timopheevi* cytoplasm, which may negatively contribute to yield. Hybrid wheat based on a chromosomal male sterile will involve *T. aestivum* cytoplasm. Another advantage of the chromosomal system is that the genetic modifications for both the sterility and the fertility restoration are in the female parent. As the male parent is normal in all respects, one female parent can easily be tested with many male parents for levels of heterosis. Greater flexibility of the male parent allows greater scope in disease resistance gene management, in that different genes can be incorporated in the male parent for different years. Even though it may be possible to isolate homozygotes with equal yield to the hybrid, this involves an uncertainty of detection and considerable time. However, the genetic components for a chromosomal system have not as yet been isolated.

Cell and tissue culture: potentials for plant breeding

By D. R. Davies

John Innes Institute, Colney Lane, Norwich NR4 7UH, U.K.

For the plant breeder, one of the objectives of cell culture systems should be their exploitation for the induction and isolation of mutant cells, which can then be regenerated as mutant plants. While a number of mutations have been recognized in plant cells *in vitro*, few have had any significance for plant breeding. There are currently a number of constraints to the exploitation of this technology, some of which are related to methodological limitations; these are likely to be overcome, but others, which relate to the nature of the attributes that the plant breeder seeks to modify, are much more intractable. There is scope for exploiting cell cutures as genetic tools, as has already been done with animal cell cultures. In contrast, the culture of organized tissues in the form of meristems or small shoots has begun to be useful a technique for plant breeders, and examples of diverse applications will be discussed. Most exploit the rapid rates of multiplication, and the assured health status of the propagules, that can be attained in culture; there is also the possibility of manipulating the genotype of these tissues. Finally, organ culture, and it is the culture of embryos that is of most interest to the plant breeder in this context, is considered; the value of embryo cultuie as a means of producing novel interspecific and intergeneric hybrids is well recognized. In addition, cultured embryos can be used as experimental systems for studying the biochemistry and molecular biology of storage product synthesis and accumulation.

The totipotency of plant cells and the relative ease with which they can be cultured *in vitro* have engendered a degree of optimism that cell and tissue culture can provide a useful new technology for plant breeders, but in only a few instances and only for particular kinds of application has this optimism been justified. I shall discuss some of the achievements as well as the limitations of cell tissue culture, excluding from consideration pollen and protoplast cultures, as these topics are discussed elsewhere in this symposium (by Hermsen & Ramanna and by Cocking), and deal first with single-cell culture, then culture of cell groups, of organized tissues and finally of plant organs.

Cell culture

One of the objectives of cell culture systems in a plant breeding context is the induction and isolation of mutant forms and the regeneration of plants from such mutants. Single-cell culture has been successful in few instances; the production of embryoids from single carrot cells was noted by Steward *et al.* (1958), of plantlets from single cells of *Macleaya cordata* by Kohlenbach (1965), and a few other examples are known (see Narayanaswamy 1977). From an analytical point there would be advantages in being able to generate embryoids or colonies from single cells, preferably plated at low densities; there also could be other benefits in avoiding the mixture of cells of differing genotype that can occur within a group, since in these circumstances a mutant can be swamped by faster-growing wild-type cells surrounding it, or be killed by the lytic products of dying cells around it.

The culture of groups of cells growing as callus masses in liquid or on solid media has formed the basis of most of the experimental work that has been undertaken, and the cells of a very large number of plant species can now be cultured in this manner. This system suffers from the cells' need to be grown at high densities, their tendency to genetic instability and the difficulty of regenerating plants at high frequencies. The two last problems may well be a function of the cell type that tends to occur in many callus cultures; this often consists of large highly vacuolated cells, whereas it is a general experience that small highly meristematic cells are less prone to these problems. There is an important area of research in the role of cell geometry, of the cell wall and of cell–cell relations in genetic stability and differentiation in plant cells. However, the technical problems of genetical stability and ease of differentiation will be overcome as more appropriate media and conditions are defined. It is a moot point whether another technical limitation, the inability to replica-plate plant cells, will be overcome.

Another kind of limitation has been noted in some experiments: the differential expression of characters in cells grown *in vitro* and those *in vivo*. For example, Widholm (1980) recently quoted four examples in which aspects of amino acid biosynthesis were different in cultured cells from those in the plants from which they were derived, or the plants to which they gave rise. One example was that of 5-methyltryptophan-resistant tobacco cells that had an altered anthranilate synthetase; plants regenerated from resistant lines selected *in vitro* did not show the altered enzyme, although cultures derived yet again from these plants once more had the modified enzyme. This limitation is certainly not ubiquitous, and examples of consistency of expression in cells *in vitro* and *in vivo* are known (see Maliga (1978) for references). Examples are nicotine content (Kinnersley & Dougall 1980), resistance to valine (Bourgin 1978), to picloram (Chaleff & Parsons 1978), to 5-bromodeoxyuridine (Márton & Maliga 1975) and to the fungal pathogen *Phytophthora parasitica* in tobacco (Helgeson *et al.* 1976) and *P. infestans* in potato (Behnke 1980). Two well documented examples of the exploitation of cell culture of significance to plant breeding are those in which *Pseudomonas tabaci* resistant tobacco strains (Carlson 1973) and *Helminthosporium maydis* resistant forms of *Zea mays* (Gengenbach *et al.* 1977) were produced. Attempts are also being made to select in culture for altered levels of specific constituents of crop plants, those of nicotine in tobacco (Collins & Legg 1979) and of urease in soybean (Polacco 1976) being good examples.

Although there are a few examples of potentially useful classes of mutants being selected, there are severe limitations on the kinds of selective techniques that can be exploited and thus of mutant forms that can be recognized *in vitro*. The complexity of many of the attributes that we seek to improve in our crops, and our ignorance of that which underlies them at a cellular and biochemical level, are severe constraints in this context. For example, while we can isolate disease-resistant cell lines when this is based on resistance to a pathogen-produced toxin (see Earle 1978), in the vast majority of instances we have insufficient understanding of that which underlies resistance to allow us to derive a selection régime. Such limitations to our understanding of the molecular biology of the components of plant productivity are likely to be greater barriers to progress in the near future than the technical problems of cell culture; the same is true of the techniques of genetic engineering in plants.

For those attributes that are only expressed in particular differentiated organized tissues, and they currently constitute a substantial proportion of those in which the plant breeder is interested, cell culture can at present offer us little. It has been considered by some that one exception to this might be storage proteins; it is assumed that they are only synthesized in seed. In these proteins, particular amino acids may be deficient, e.g. lysine in barley and rice, and methionine in legumes.

It has been suggested that if mutant cells could be isolated in culture that overproduce the required amino acid, it is possible that the seed proteins might also contain more of that particular amino acid. This is a tenuous argument, and Boulter & Crocomo (1979) have stated that in legumes there is as yet no evidence that protein quality is dependent on the supply of particular amino acids. Chaleff & Carlson (1975) isolated mutant cultures of rice, on the basis of their resistance to the lysine analogue S-β-aminoethylcysteine, that overproduced lysine, but as no plants could be differentiated from the cultures the consequence of these mutational changes on the seed protein was not tested. More recently, Hibberd *et al.* (1980) have generated *Zea mays* lines able to grow on normally inhibitory levels of lysine plus threonine and found that the selected cultures had increased concentrations of particular amino acids; for example, in one line the lysine was twice, and in another the methionine concentration 3.8 times, as high as in the control. Only free aspartate-derived amino acids were increased. Plants were regenerated from these cultures, but no further generations could be derived and so once again the consequences in terms of seed protein could not be evaluated.

One of the problems of callus cultures, their proneness to genetic instability, has been turned to advantage in sugar cane, where variants have proved useful (Nickell 1977). Forms showing improved resistance to a virus and to fungal pathogens have been obtained; it has to be noted, however, that sugar cane is a vegetatively propagated species, with very high chromosome numbers and tolerant of chromosomal changes.

It is difficult to summarize a topic that has generated on average one symposium per year in the last few years, but while a great deal of attention has been drawn to the potential role of this technology in plant breeding, the achievements are minimal as yet. The extent to which this will alter will depend in part on our ability to improve the technology of cell culture and attain a greater expression of characters *in vitro* than is now possible. Cell cultures have not yet been exploited as genetic tools in plants, for mapping, and for complementation and linkage studies, as they have been in human cell systems, for example (McKusick & Ruddle 1977). Neither have we examined whether it is feasible to induce, and if so what might be the consequences of duplicating, certain chromosomal regions in plant cells grown *in vitro*. In mouse cell cultures, lines can be selected that show enhanced resistance to the antifolate drug methotrexate, due to increased dihydrofolate reductase activity. The resistant cells achieve this by a selective amplification of the genes for dihydrofolate reductase (Alt *et al.* 1978). DNA amplification occurs in plants *in vivo* in the well documented example of the giant chromosomes of *Nicotiana* hybrids (Gerstel & Burns 1976). Claims of DNA amplification in plant cells grown *in vitro* have been made (see Buiatti 1977), but its phenotypic consequences have not been analysed; this is an important challenge for us. It is significant that in the mouse cells, DNA other than the gene sequences directly selected was amplified (Nunberg *et al.* 1978). If that also occurred in plant cells, then genes that could not themselves be directly selected in culture, but which were closely linked to those that could, might be amplified and the consequences examined.

The culture of tissues

The ability to culture organized tissues in the form of very small shoots or meristems has allowed a most valuable application of plant tissue culture. Meristem culture has long been used for the production of virus-free plants, while the culture of small shoots and meristems is being exploited for the rapid vegetative multiplication (micropropagation) of a range of horticultural and agri-

cultural plants (Holdgate 1977; Murashige 1978). In the context of plant breeding it is now also possible to cite many examples in which it is advantageous to exploit either the ability to propagate by tissue culture genotypes in which there is no natural or simple method of vegetative propagation, or the more rapid rates of multiplication that can be attained *in vitro*.
Such applications include the following.

(*a*) New varieties are often not available to the agricultural or horticultural industry for many years after the recognition of their value, simply because of the time taken to generate appropriate quantities for large-scale planting. Examples among vegetatively propagated horticultural crops are daffodils, freesia, gladioli and alstromeria, in all of which the natural rates of multiplication are low, and 10 or more years may elapse before a desired strain becomes available. In all of these micropropagation techniques are being used to speed up the release of new varieties to the industry. Another example in which the technique has been exploited is in the multiplication of new strains of rootstocks for top fruit.

(*b*) There is a need to maintain new varieties of vegetatively propagated plants in a disease-free condition for as long as possible during their period of multiplication before release. This means that micropropagation can be an attractive and economic alternative to conventional methods of multiplication even though a species may have a rapid rate of natural propagation. We may well find in the near future that strawberries and potatoes will be in this category.

(*c*) The breeding system of a crop plant can impose limitations on the multiplication of a genotype. In some such instances, micropropagation techniques can be a useful means of overcoming this, as the following examples illustrate.

(i) *Incompatibility systems.* In particular forms of *Brassica oleracea*, F_1 hybrids have a considerable commercial attraction; their production is dependent on the selection and maintenance of pairs of inbred parents. The sporophytic incompatibility system of these genotypes means that maintenance of the parents depends on bud pollination, which is both difficult and expensive; micropropagation offers an alternative method of maintenance and multiplication (Dunwell & Davies 1975).

(ii) *Male sterility.* Maintenance of male-sterile genotypes demands a continuous process of backcrossing, and again micropropagation can be an attractive alternative. For example, male-sterile onions can be readily propagated in tissue culture (Hussey 1978), and several male-sterile lines of wheat have been multiplied in this manner (G. Hussey, personal communication).

(iii) *Dioecious forms.* Tissue culture has been used for the clonal multiplication of selected genotypes of asparagus that are required as parental plants for the production of commercial quantities of seed (Dore 1975).

(iv) *Heterozygous genotypes.* The multiplication of large quantities of particular heterozygous genotypes to be used as parents in a seed production programme can now be achieved even though no, or only a slow, method of natural vegetative propagation exists. The onion crop again provides one example of such a species, and both diploid and tetraploid sugar beet (Hussey & Hepher 1978) may be other candidates for incorporating a micropropagation step into the breeding programme.

(*d*) Multiplication of existing superior genotypes. In some species the long life cycle or/and the heterozygosity of the plants renders conventional breeding methodology difficult; in addition, in some such instances there is no natural method of vegetative propagation. The availability of a micropropagation technique that would allow existing superior genotypes to be cloned could significantly improve the average level of performance of such a crop; a prime example in which clonal plantings of superior genotypes will lead to an increase in productivity is the oil palm (Jones 1974).

(e) *Correlation of seedling and mature plant responses.* Selection for disease resistance is often facilitated if it can be based on the screening of seedlings. However, their responses need not be identical with those of mature plants, and the extent to which they are correlated is not easy to establish in many instances. In homozygous forms, sister plants can be used for this comparison, but with heterozygotes this is not possible. By exploiting micropropagation, my colleagues Matthews & Dunwell (1979) were able to overcome this problem in the carnation crop; from a given seedling, the tip was taken for culture and used to generate a clone of adult plants. The response of the remainder of the seedling to a given pathogen and of the adult plant derived from the shoot tips could then be compared.

(f) Manipulation of the genotype.

(i) *Production of polyploid forms.* Treatment of cultures with colchicine has allowed higher rates of production of polyploids than is usually possible by more conventional techniques. In one series of experiments with freesia, which involved placing the cultures in a colchicine solution for 24 h, 27% of the plants regenerated from a treated diploid culture became tetraploid, whereas none were found in the control (Davies 1973). High yields of tetraploid carnations have been produced in a comparable manner (Dunwell & Cornish 1978).

(ii) *Manipulation of cytoplasmic male sterility.* It has been reported that cytoplasmic male sterility in sugar beet can be 'cured' by heat treatment (Lichter 1978) and that it is graft transmissible (Curtis 1967). Sugar beet is highly heterozygous and is not readily propagated by vegetative means, but the availability of clonal material would greatly facilitate such experimental approaches. Furthermore, heat treatment of cultured tissues can be readily undertaken, as well as the grafting of propagules *in vitro*. The feasibility of manipulating male sterility in this way is currently being examined by using tissue culture, and the provision of clonal material is also aiding the comparative analysis of mitochondrial DNA in cytoplasmic male sterile and in fertile genotypes (A. Powling, personal communication).

(iii) *Analysis of genotroph induction.* The production of genotrophs in flax (Durrant 1962) is accompanied by numerous changes in the genome (Cullis 1977). The study of one of these changes, that induced in the ribosomal genes, has been facilitated by the availability of a tissue culture system (Cullis & Charlton 1981). The terminal portion of the shoot of young flax seedlings was harvested at various times after initiating the treatments that induce the genotrophs, and the ribosomal DNA (rDNA) within them assayed; the remainder of the stem below this region was then cultured, and from each of the axillary meristems that were subsequently induced to develop, the rDNA was extracted and assayed. In this way Cullis & Charlton could determine when the changes occurred in the rDNA during the process of induction. They showed that the changes occurred rapidly and only in the terminal regions of the stem, an analysis that would otherwise be extremely difficult to achieve.

(iv) *Induction of mutations.* Adventitious meristems can develop from single cells; if these cells have been modified by exposure to mutagenic agents, wholly mutant meristems and plants may be immediately generated. In such meristems the competition that occurs between wild-type and mutant cells, which can result in the suppression of the latter, is avoided. The end result should be a higher rate of recovery of mutant plants. Adventitious buds are readily generated in culture, and as Broertjes *et al.* (1976) have shown with chrysanthemum, they can be a useful source of induced mutations.

(g) *Storage of genotypes.* The maintenance of selected heterozygous lines to be used as parents in the production of commercial varieties can often be difficult. If they are maintained by seed multiplication, there is a danger of the occurrence of genetic drift. Storage of vegetatively propagated material can equally be difficult. Tissue cultures can, however, be stored for many months and even years in some instances, simply by keeping them on a nutrient agar medium at 4 °C and at low light intensities. This offers an easy, cheap and, furthermore, disease-free system of maintaining parental lines for variety production.

(*h*) Provision of disease-free material. Disease-free strains are required for an unbiased evaluation of potential new varieties and also for the selection of parents. Dale (1975) has suggested that the breeding of certain grass species in which virus infection rapidly leads to a marked reduction in plant vigour would be aided by the availability of disease-free forms.

These examples illustrate the opportunities that are already available to the plant breeder to exploit the culture of organized tissues, but a further expansion of the technology will undoubtedly occur as the culture of a wider range of species and genera becomes possible, and as plant breeders recognize the role it can play.

Organ culture

For the plant breeder it is the culture of embryos that is of primary interest in this context. The production of many interspecific hybrids has been possible as the result of the rescue of immature embryos by excision from the ovary and their subsequent culture *in vitro*. A recent review (Raghavan 1977) has summarized the applications of embryo culture and the range of hybrids produced in this manner. Included among them are intergeneric hybrids involving *Triticum* and related genera, and *Zea mays* and its relatives, as well as interspecific crosses involving *Triticum*, *Hordeum*, *Oryza*, *Sorghum*, *Nicotiana* and *Hordeum*, to list but a few of many important crop plants. While it is unlikely that such hybrids themselves or their allopolyploid derivatives will be useful as crop plants, they do offer a means of achieving an interspecific or intergeneric transfer of chromosomes or of chromosome fragments. For this latter purpose they can sometimes offer an easier alternative to the route offered by protoplast fusion. The interspecific and intergeneric embryos have the advantage that nuclear fusion, cell wall formation and initial cell divisions have already been achieved. The elimination of particular chromosomes in interspecific embryos can occur naturally, as in the hybrids of *Hordeum vulgare* and *H. bulbosum*, and of *Nicotiana plumbaginifolia* and *N. tabacum*, leading in these instances to the production of offspring in which only one parental genome is present; this has been the basis of the large-scale production by embryo culture of barley haploids (Kasha 1974). Until more effort is devoted to establishing whether the phenomenon of chromosome elimination occurs in other interspecific and intergeneric embryos, it is impossible to speculate on the wider applicability of this technology of producing haploids. A method of attaining an interspecific transfer of chromosome fragments has been developed by Pandey (1980), in which pollen is exposed to high doses of ionizing radiation to fragment the chromosomes, before its use for pollination. Fragments of the paternal chromosomes are then incorporated into, or attached to, the maternal chromosomes within the embryo. These experiments have not involved embryo culture, and few progeny incorporating alien chromosome fragments have been produced. My colleague, J. M. Dunwell, is attempting to modify Pandey's approach by inducing a proliferation of the cultured embryos produced after interspecific or intergeneric pollination with irradiated pollen; in this way he hopes to generate from each embryo a number of derivatives, each of which will have the maternal genome and also a different paternal fragment(s). This may enable us to sample a greater range of the fragments that can be included in the embryos.

Few attempts have been made to culture plant embryos for experimental purposes other than the generation of hybrids. This is in marked contrast to animal embryology, which has a long tradition of experimental work. In many crop plants the embryo is the economically important component, yet we know singularly little about it in cellular, biochemical or molecular terms. Embryo culture could be useful in this respect and I shall describe some of our own work on peas (*Pisum sativum*) to illustrate this. Culture methods have improved to an extent that comparable growth rates can be achieved *in vivo* and *in vitro* in *Phaseolus vulgaris* (Thompson *et al.* 1977) and in peas (Stafford & Davies 1979), although in both species the period of growth over which these rates can be maintained is still relatively limited. Mature legume seed is composed almost entirely of the embryo, with the swollen cotyledons composing the storage tissue. The number and size of cells within the cotyledons are the determinants of seed size, and while genetic variation exists for both components (Davies 1975, 1977) we do not have at present much knowledge of their respective importance. The DNA in the cotyledon cells is highly endoreduplicated, the extent of DNA duplication being proportional to cell size (Davies 1977). By an appropriate culture technique we can trigger these cells, which normally remain in interphase, into a prophase stage, in which giant polytene-like chromosomes are seen (Marks & Davies 1979). Embryo culture could be used to examine the way in which factors influencing seed size affect the two components, cell size and cell number. It could also be used to study the control of storage product accumulation within the seed. We have shown that comparable amounts of protein and starch are synthesized *in vitro* and *in vivo* in peas (Stafford & Davies 1979). Beyond this we have examined the synthesis of legumin, one of the two main storage proteins in peas, in cultured embryos, and compared it with that occurring *in vivo*. Earlier work had suggested that legumin was synthesized at a fairly late stage of development of the embryo, when the greater proportion of the cells of the cotyledon were becoming endoreduplicated; secondly, it was believed that legumin synthesis could not be initiated in cultured embryos (Millerd *et al.* 1975). By using a more sensitive assay for legumin, an enzyme-linked immunoabsorbent assay (ELISA), which allows us to detect nanogram quantities of the protein (Domoney *et al.* 1980), we have shown that legumin is synthesized in much younger embryos than hitherto assumed, an observation to which I will return later. Secondly, with improved culture techniques we have demonstrated that legumin synthesis can be initiated *in vitro* (Domoney *et al.* 1980).

Returning to the observation of the presence of legumin in the cells of very young embryos, this implies either that there is a low rate of synthesis even in the diploid cells of the young cotyledon or that there are already a few endoreduplicated cells present, and it is these that are synthesizing the protein. Should the former be true, it is important to test whether other diploid cells within the plant, and even cells in culture, can synthesize legumin, albeit at a very low level, but levels that we might now detect with the ELISA technique. It has been suggested that callus cells derived from cultures of *Vicia* cotyledons can synthesize low levels of storage protein (Muntz, quoted in Boulter & Crocomo 1979). The possibility of selecting mutant cells in culture that can overproduce particular storage proteins is attractive, and the aim in peas would be to enhance the production of legumin, which has a higher proportion of sulphur amino acids than some of the other seed proteins.

Embryo culture is being used also for studying another aspect of storage product synthesis in peas. It has been recently shown (Davies 1980) that there are mutants in peas, somewhat akin to those in maize and barley, in which both protein and carbohydrate composition is altered within the seed. In peas the two seed phenotypes, round and wrinkled, differ in starch quantity and quality, sugar content and storage protein composition, the proportion of legumin being higher in round seed (Davies 1980). The nature of the metabolic changes induced by the allelic alternatives at the r_a locus, which is involved in the determination of these two phenotypes, is not known, but we are using embryo culture to examine the proteins synthesized by these two genotypes when grown under various conditions to try to analyse the relation of carbohydrate and protein synthesis, and how it may be manipulated. We therefore have mutants in peas that affect storage product composition, we can culture the embryos in which these products are synthesized and stored, and we can define the ways in which we need to improve the phenotype. An important limitation, however, is the dearth of knowledge of the biochemistry and molecular biology of these important components of economic yield – the seed – and this needs to be remedied if we are to successfully manipulate and modify them by plant breeding.

Conclusion

A cautious optimism may be an appropriate conclusion; after all, plant breeders and geneticists took little interest in cell, tissue and organ culture until recently, but already the value of meristem and shoot culture is widely recognized. Other aspects of the subject will prove attractive as techniques improve and new applications are recognized, and if genetic engineering is to contribute to plant improvement, it will be mediated to a substantial extent through the manipulation of cells in culture.

References (Davies)

Alt, F. W., Kellems, R. E., Bertino, J. R. & Schimke, R. T. 1978 Selective multiplication of dihydrofolate reductase genes in methotrexate-resistant variants of cultured murine cells. *J. biol. Chem.* **253**, 1357–1370.

Behnke, M. 1980 General resistance to late blight of *Solanum tuberosum* plants regenerated from callus resistant to culture filtrates of *Phytophthora infestans*. *Theor. appl. Genet.* **56**, 151–152.

Boulter, D. & Crocomo, O. J. 1979 Plant cell culture implications: legumes. In *Plant cell and tissue culture* (ed. W. R. Sharp *et al.*), pp. 615–631. Ohio State University Press.

Bourgin, J.-P. 1978 Valine-resistant plants from *in vitro* selected tobacco cells. *Molec. gen. Genet.* **161**, 225–230.

Broertjes, C., Roest, S. & Bokelmann, G. S. 1976 Mutation breeding of *Chrysanthemum morifolium* Ram. using *in vivo* and *in vitro* adventitious bud techniques. *Euphytica* **25**, 11–19.

Buiatti, M. 1977 DNA amplification and tissue culture. In *Plant cell, tissue, and organ culture* (ed. J. Reinert & Y. P. S. Bajaj), pp. 358–374. Berlin: Springer-Verlag.

Carlson, P. S. 1973 Methionine sulfoximine-resistant mutants of tobacco. *Science, N.Y.* **180**, 1366–1368.

Chaleff, R. S. & Carlson, P. S. 1975 Higher plant cells as experimental organisms. In *Modification of the information content of plant cells (2nd John Innes Symposium)* (ed. R. Markham *et al.*), pp. 197–214, North Holland/American Elsevier.

Chaleff, R. S. & Parsons, M. F. 1978 Direct selection *in vitro* for herbicide-resistant mutants of *Nicotiana tabacum*. *Proc. natn. Acad. Sci. U.S.A.* **75**, 5104–5107.

Collins, G. B. & Legg, P. D. 1979 Use of tissue and cell culture methods in tobacco improvement. In *Plant cell and tissue culture* (ed. W. R. Sharp *et al.*), pp. 585–614. Ohio State University Press.

Cullis, C. A. 1977 Molecular aspects of the environmental induction of heritable changes in flax. *Heredity, Lond.* **38**, 129–154.

Cullis, C. A. & Charlton, L. 1981 The induction of ribosomal DNA changes in flax. *Pl. Sci. Lett.* **20**, 213–217.

Curtis, G. 1967 Graft-transmission of male sterility in sugar beet (*Beta vulgaris* L.) *Euphytica* **16**, 419–424.

Dale, P. J. 1975 Tissue culture in plant breeding. *Rept Welsh Pl. Breed. Stn 1975*, pp. 101–115.

Davies, D. R. 1973 *In vitro* propagation of *Freesia*. *John Innes A. Rept*, pp. 67–68.

Davies, D. R. 1975 Studies of seed development in *Pisum sativum*. 1. Seed size in reciprocal crosses. *Planta* **124**, 297–302.

Davies, D. R. 1977 DNA contents and cell number in relation to seed size in the genus *Vicia. Heredity, Lond.* **39**, 153–163.

Davies, D. R. 1980 The r_a locus and legumin synthesis in *Pisum sativum. Biochem. Genet.* **18**, 167–179.

Domoney, C., Davies, D. R. & Casey, R. 1980 The initiation of legumin synthesis in immature embryos of *Pisum sativum* L. grown *in vivo* and *in vitro*. *Planta* **149**, 454–460.

Dore, C. 1975 La multiplication clonale de l'asperge (*Asparagus officinalis* L.) par culture *in vitro*: son utilization en selection. *Annls Amél. Pl.* **25**, 201–224.

Dunwell, J. M. & Cornish, M. 1978 Production of tetraploids in carnation. *John Innes A. Rept*, p. 42.

Dunwell, J. M. & Davies, D. R. 1975 Technique for propagating inbred brussel sprout strains by means of tissue culture. *Grower* **84**, 105–106.

Durrant, A. 1962 The environmental induction of heritable change in *Linum. Heredity, Lond.* **17**, 27–61.

Earle, E. D. 1978 Phytotoxin studies with plant cells and protoplasts. In *Frontiers of plant tissue culture* (ed. T. A. Thorpe), pp. 363–372. The International Association for Plant Tissue Culture.

Gengenbach, B. G., Green, C. F. & Donovan, C. M. 1977 Inheritance of selected pathotoxin resistance in maize plants regenerated from cell cultures. *Proc. natn. Acad. Sci. U.S.A.* **74**, 5113–5117.

Gerstel, D. U. & Burns, J. A. 1976 Enlarged euchromatic chromosomes (megachromosomes) in hybrids between *Nicotiana tabacum* and *N. plumbaginifolia. Genetica* **46**, 139–153.

Helgeson, J. P., Haberlach, G. T. & Upper, C. D. 1976 A dominant gene conferring disease resistance to tobacco plants is expressed in tissue cultures. *Phytopathology* **66**, 91–96.

Hibberd, K. A., Walter, T., Green, C. E. & Gegenbach, B. C. 1980 Selection and characterization of a feedback-insensitive tissue culture of maize. *Planta* **148**, 183–187.

Holdgate, D. P. 1977 Propagation of ornamentals by tissue culture. In *Plant cell, tissue, and organ culture* (ed. J. Reinert & Y. P. S. Bajaj), pp. 18–43. Berlin: Springer-Verlag.

Hussey, G. 1978 *In vitro* propagation of the onion *Allium cepa* by axillary and adventitious shoot proliferation. *Scient. Hort.* **9**, 227–236.

Hussey, G. & Hepher, A. 1978 Clonal propagation of sugar beet plants and the formation of polyploids by tissue culture. *Ann. Bot.* **42**, 477–479.

Jones, L. H. 1974 Propagation of clonal oil palm by tissue culture. *Oil Palm News* **17**, 1–8.

Kasha, K. J. 1974 Haploids from somatic cells. In *Haploids in higher plants* (ed. K. J. Kasha), pp. 67–87. University of Guelph Press.

Kinnersley, A. M. & Dougall, D. K. 1980 Correlation between the nicotine content of tobacco plants and callus cultures. *Planta* **149**, 205–206.

Kohlenbach, H. W. 1965 Über organisierte Bilgungen aus *Macleaya cordata* Kallus. *Planta* **64**, 37–40.

Lichter, R. 1978 The restoration of male fertility in cytoplasmic male-sterile sugar beet by heat treatment and meristem culture. *Z. PflZücht.* **81**, 159–165.

McKusick, V. A. & Ruddle, F. H. 1977 The status of the gene map of the human chromosomes. *Science, N.Y.* **196**, 390–405.

Maliga, P. 1978 Resistance mutants and their use in genetic manipulation. *Frontiers of plant tissue culture* (ed. T. A. Thorpe), pp. 381–392. The International Association for Plant Tissue Culture.

Marks, G. E. & Davies, D. R. 1979 The cytology of cotyledon cells and the induction of giant polytene chromosomes in *Pisum sativum. Protoplasma* **101**, 73–80.

Márton, L. & Maliga, P. 1975 Control of resistance in tobacco cells to 5-bromodeoxyuridine by a simple mendelian factor. *Pl. Sci. Lett.* **5**, 77–81.

Matthews, P. & Dunwell, J. M. 1979 The relationship between the resistance of seedlings and cuttings to Fusarium wilt in the carnation. *John Innes A. Rept*, p. 48.

Millerd, A., Spencer, D., Dudman, W. F. & Stiller, M. 1975 Growth of immature pea cotyledons in culture. *Aust. J. Pl. Physiol.* **2**, 51–59.

Murashige, T. 1978 The impact of plant tissue culture on agriculture. In *Frontiers of plant tissue culture* (ed. T. A. Thorpe), pp. 15–26. The International Association for Plant Tissue Culture.

Narayanaswamy, S. 1977 Regeneration of plants from tissue culture. In *Plant cell, tissue, and organ culture* (ed. J. Reinert & Y. P. S. Bajaj), pp. 179–206. Berlin: Springer-Verlag.

Nickell, L. G. 1977 Crop improvement in sugarcane: studies using *in vitro* methods. *Crop Sci.* **27**, 717–719.

Nunberg, J. H., Kaufman, R. J., Schimke, R. T., Urlaub, G. & Chasin, L. A. 1978 Amplified dihydrofolate reductase genes are localized to a homogeneously staining region of a single chromosome in a methotrexate-resistant Chinese hamster ovary cell line. *Proc. natn. Acad. Sci. U.S.A.* **75**, 5553–5556.

Pandey, K. K. 1980 Further evidence for egg transformation in Nicotiana. *Heredity, Lond.* **45**, 15–29.

Polacco, J. C. 1976 Nitrogen metabolism in soybean tissue culture. *Pl. Physiol.* **58**, 350–357.

Raghavan, V. 1977 Applied aspects of embryo culture. In *Applied and fundamental aspects of plant cell, tissue, and organ culture* (ed. J. Reinert & Y. P. S. Bajaj), pp. 374–397. Berlin: Springer-Verlag.

Stafford, A. & Davies, D. R. 1979 The culture of immature pea embryos. *Ann. Bot.* **44**, 315–321.

Steward, F. C., Mapes, M. O. & Mears, K. 1958 Growth and organized development of cultured cells. *Am. J. Bot.* **45**, 705–708.

Thompson, J. F., Madison, J. T. & Meunster, A.-M. E. 1977 *In vitro* culture of immature cotyledons of soya bean (*Glycine max* L. Merr.). *Ann. Bot.* **41**, 29–39.

Widholm, J. 1980 Differential expression of amino acid biosynthetic control iso-enzymes in plants and cultured cells. In *Plant cell cultures: results and perspectives* (ed. F. Sala *et al.*), pp. 157–159. Elsevier/North Holland.

Discussion

S. BRIGHT (*Rothamsted Experimental Station, Harpenden, Herts, U.K.*). Some recent work on plant mutants that accumulate amino acids is relevant here after Professor Davies's paper in which he questioned whether accumulation would be expressed in seeds. Complex diploid tissue cultures of maize (Hibberd *et al.* 1980) or mature embryos of diploid barley have been used to select mutants resistant to lysine plus threonine (Bright *et al.* 1981). In both cases, threonine and methionine are accumulated in the growing tissues. One barley mutant, Rothamsted 2501, contains a single dominant gene for resistance (associated with recessive lethality). Normal seeds contain little soluble threonine (less than 1% of total threonine) whereas seeds from resistant plants have 15% of the total threonine in the soluble fraction. This is sufficient to change the total threonine content. There is evidence for increased total methionine also in the barley mutant described above: as soluble methionine is very low this must be in protein.

Reference

Bright, S. W. J., Miflin, B. J. & Rognes, S. E. 1981 Threonine accumulation in the seeds of a barley mutant with an altered aspartate kinase. (Manuscript in preparation.)

Opportunities from the use of protoplasts

By E. C. Cocking

*Plant Genetic Manipulation Group, Department of Botany, University of Nottingham,
Nottingham NG7 2RD, U.K.*

Plant protoplasts of several horticultural and crop species can now be readily regenerated into plants. There are now available several opportunities for their use in the manipulation of genetic systems in plant breeding. Protoplast cloning has recently been shown to produce additional genetic variation in potatoes; the possibility is examined for protoplast cloning of seed producing crop species for new genetic variation. Fusion of protoplasts of different species is now providing an additional method of hybridization; sexually incompatible species can be hybridized and horticulturally useful hybrids are now being produced. Many possibilities exist for hybridization assessments, both nuclear and cytoplasmic, between various crop species; however, the extent to which these wider hybridizations will produce useful genetic variation is not yet clear, and in many instances plant regeneration from these cultured cell hybrids is not yet possible.

Plant protoplasts are also providing an opportunity for the transfer of genes between different species. This may be by fusion with an irradiated protoplast system, or by direct transformation. Transfer of genes by using *Agrobacterium* plasmid as a vector system appears promising, and fusions with wild-type protoplasts will ensure the regeneration of non-tumorous plants.

1. Introduction

At this Discussion Meeting it will be most useful to clarify what has already been accomplished with the use of protoplasts, to indicate the main obstacles impeding further developments and to forecast the most likely future applications of this new technology to plant breeding. Comprehensive earlier assessments of the role of plant tissue culture and protoplasts have already been provided (Riley 1979; Cocking & Riley 1981), but recent advances in various aspects of protoplast manipulation are providing additional results and new vistas.

Because protoplasts are single cells they provide opportunities to clone plants at the single cell level. Also, because the cell wall is absent they can be fused with other protoplasts, thereby providing opportunities for somatic hybridization. The absence of the cell wall also facilitates the uptake of DNA enabling evaluations of transformation to be carried out at a level comparable with those of microbiological transformations. These manipulations are not restricted to the cellular level because sometimes whole plants can be regenerated by suitable tissue culture procedures, and then subjected to a conventional breeding programme.

In this survey of the opportunities in plant breeding from the use of such protoplast systems, it will be assumed that the reader is familiar with the techniques involved; these have been fully described previously (Evans & Cocking 1974), and more recently in the practical handbook from this laboratory (Power & Davey 1979).

2. Cloning of plants from protoplasts

Phenotypic diversity appears to be a ubiquitous occurrence in cell culture, which is manifested among plants regenerated from cell culture. As recently discussed by Scowcroft & Larkin (1980), there is reason to suspect a genetic, or at least an epigenetic, basis for this variation. As also emphasized by these investigators, plant tissue culture *per se* appears to be an unexpectedly rich and novel source of genetic variation, which is already being utilized in plant improvement. All of the major crop species are included among the 300 or so species from which plants have been regenerated from cell cultures, but not necessarily from protoplasts.

TABLE 1. SOMACLONAL VARIATION IN *SOLANUM TUBEROSUM*

(From Scowcroft & Larkin (1980).)

source material	number of somaclones screened	characters displaying enhanced phenotypic variation	reference
Leaf mesophyll protoplasts (Russet Burbank)	> 10 000	tuber shape, yield, maturity date, photoperiod requirement, plant morphology, enhanced resistance to *Alternaria solani* and *Phytophthora infestans*	Shepard *et al.* (1980)
Suspension culture protoplasts of a dihaploid line	< 211	tuber shape	Wenzel *et al.* (1979)

The ability to dissect whole plants into single cells, through the use of protoplasts, has now stimulated a considerable effort to find out whether such genetic variation can be enhanced by protoplast cloning. Procedures such as protoplast cloning, which have the objective of enhancing a popular variety rather than creating a new one, are particularly attractive to the plant breeder for obvious reasons. This approach has been pioneered by Shepard and his collaborators, particularly for potatoes and more recently for other vegetatively propagated species. Shepard *et al.* (1980) have, however, cautioned regarding the interpretation of the results relating to improvement in any one character as a result of protoplast cloning (table 1). In any such breeding programme, clones with an acknowledged improvement in one character must be exposed to adequate environmental pressures (both physical and biological) to eliminate any that are simultaneously deficient in some critical feature. Nevertheless, the results obtained so far have been sufficiently encouraging to convince plant breeders in several countries of the need to evaluate their own potato varieties in this respect. The need for evaluations to be performed on varieties suitable for particular countries highlights the need for the procedures for protoplast isolation, culture, division and regeneration into whole plants to be applicable to a very wide range of potato varieties. Plant breeders in the U.K. will rapidly lose interest if the protoplast cloning procedures are only applicable to the Russet Burbank variety. Fortunately this does not appear to be the situation. Several U.K. recommended varieties respond well under the sophisticated growth and cultural conditions developed by Shepard and his collaborators (P. Day, personal communication). It should be noted that Wenzel *et al.* (1979) obtained a rather high uniformity of plants regenerated from their potato mesophyll protoplasts. This suggests that different varieties of potato may possess markedly different levels of inherent genetic variation.

Meaningful evaluation of protoplast cloning for potato improvement depends on reproducible regeneration of plants from protoplasts and the field testing of the plants obtained. Recurrent selection from clonal material may further enhance selection. With potato, and with any other species subject to this new breeding technology, it is important to keep the cultural procedures as simple as possible; there is a need to compare the rather simple shoot tip culture isolation procedures with the far more sophisticated procedures employed by Shepard *et al.* (1980).

At present there is little evidence to suggest that this type of somaclonal variation could make a significant contribution to plant improvement of seed propagated species. However, there has been very little work comparable with that using potato protoplasts. Since, also, at present, the origin of somaclonal variation is not fully understood, even in potatoes, it is worth while to explore the situation in seed propagated species. Recently the ability of *Medicago sativa* (alfalfa, lucerne) leaf protoplasts to regenerate into plants at high frequency, under relatively simple growth conditions (Santos *et al.* 1980), has stimulated work towards such an evaluation of this important forage legume.

As discussed by Scowcroft & Larkin (1980), in polyploid species such as sugar cane and potatoes, aneuploidy could account for some variation, but the reasons for the variation may be more fundamental. Recent molecular studies of genome organization in eukaryotes has indicated that the somatic genome is not static but highly variable. Moreover, in vegetatively propagated species this variation might be enhanced by cryptic chromosomal changes such as small deletions, additions, transpositions or inversions, which would normally be screened out during meiosis. There is no doubt that somatic variation in plants will prove to be an exciting area of research in which the use of protoplasts will feature to a large extent. The regeneration of crop plants from protoplasts is still very much the acid test for regeneration capability (Bhojwani *et al.* 1977), and we still lack this capability for the cereals and grain legumes. There are indications, now that callusing and plant regeneration from cereal leaves have been obtained (Wernicke & Brettell 1980), that further work may resolve this present deficiency. In the meantime, quite apart from protoplast studies, detailed studies on the genetic variation ensuing in plants regenerated from *in vitro* cultures of our major crop species could be particularly rewarding to the plant breeder.

3. Somatic hybridization by protoplast fusion

In higher plants, just as in the fungi, the first essential step in any parasexual cycle will be heterokaryosis as a result of fusion of somatic protoplasts. If this is followed by the fusion of nuclei and the development of hybrid cells after mitosis, then this somatic hybridization component of the parasexual cycle will be complete. Moreover, if diploidization associated with the fusion process results in the formation of amphiploid hybrid plants, then protoplast fusions will be providing an alternative to gametic fusions for plant hybridization (Cocking 1979).

Practical details of the various procedures involved in protoplast isolation and fusion to produce heterokaryons have been fully described previously (Power & Davey 1979). Although there is still much current work aimed at increasing the percentage of heterokaryons that can be obtained when protoplasts of different species are fused, it is clear that fusion itself is not a major incompatibility consideration. Usually approximately 1% heterokaryon formation can be obtained, and provided the culture conditions are satisfactory, and provided an adequate selection pressure can be put on the system, hybrid callus (or hybrid embryoids) can be selected. Sometimes it is possible to select heterokaryons manually.

Perhaps it should be re-emphasized that the somatic hybridization approach need not be restricted to wide crosses. For instance, in potato breeding it has been suggested by Wenzel et al. (1979) that an ideal procedure would be to fuse protoplasts of different dihaploid hybrids. The heterozygous dihaploids would be combined to a completely heterozygous tetraploid plant, without any meiotic segregation: propagated vegetatively it could immediately become a stable variety.

Results obtained so far on the consequences of fusion of protoplasts have provided good evidence that, in many instances, fusions between those from sexually compatible species will result in the formation of heterokaryons in which both nuclear genomes are capable of forming stable amphiploid nuclear hybrids. Organogenesis from hybrid callus will then result in the regeneration of somatic hybrid plants with both sets of parental chromosomes. In our studies within the *Petunia* genus (Cocking 1979) we have produced a large percentage of amphidiploid somatic hybrid plants by fusing wild-type leaf protoplasts of one species with albino protoplasts from cell suspension cultures of the other species. This has resulted in the production of flowering plants, with 28 chromosomes, of the somatic hybrid *Petunia parodii* ($2n = 14$) $\otimes$ *Petunia hybrida* ($2n = 14$), and of *Petunia parodii* ($2n = 14$) $\otimes$ *Petunia inflata* ($2n = 14$). While *P. parodii* and *P. hybrida* can be crossed sexually, it is known that *P. parodii* and *P. inflata* cannot be crossed sexually by standard procedures, since they possess prezygotic unidirectional sexual incompatibilities.

Recently we have investigated the consequences of fusion of *Petunia parodii* leaf protoplasts with albino protoplasts from cell suspensions of *Petunia parviflora*. These two species are sexually incompatible, probably due to both prezygotic and postzygotic isolation mechanisms, and all previous attempts at sexual hybridization have failed. Our somatic protoplast fusions have enabled us to produce somatic hybrid plants with 32 chromosomes, *P. parodii* ($2n = 14$) $\otimes$ *P. parviflora* ($2n = 18$) which are amphidiploids ($4n = 32$) and some with 31 chromosomes (Power et al. 1980). Differences in size and morphology of the chromosomes of these two species have proved sufficient to enable us to establish the presence of both parental sets in the somatic hybrid nuclei. Other somatic hybrids have also been produced between species that are very difficult or impossible to hybridize conventionally, e.g. *Lycopersicon esculentum* and *Solanum tuberosum* (Melchers et al. 1978), *Datura innoxia* and *Atropa belladona* (Krumbiegel & Schieder 1979); *Arabidopsis thaliana* and *Brassica campestris* (Gleba & Hoffmann 1979).

It would seem likely that the main opportunities for the plant breeder centre on the use of protoplast fusions for hybridizations within the same genus, or between closely related genera, particularly if fertile, seed-producing somatic hybrids are desired. An example where tetraploids from somatic fusion would be useful is the *Lolium–Festuca* hybrid. In this cross the diploid, even if obtained sexually, often lacks fertility, and chromosome doubling by colchicine treatment is frequently impossible. Thus tetraploid production, somatically, between these species might produce fertile hybrids. The present main impediment to finding out if this would be so is our present inability to regenerate plants from protoplasts in these grass species. Many intergeneric hybrids are desirable in programmes of crop improvement between sexually incompatible species. A typical example within the *Vicias* is the desirability of crossing *V. faba* with *V. narbonensis* to convey chocolate spot resistance from *V. narbonensis* into the cultivated disease-sensitive *V. faba*. Progress at the protoplast culture level has already reached the stage of callus formation from protoplasts of the parental species, but regeneration of plants still remains unresolved. Numerous examples in other genera have previously been itemized

(Sanchez-Monge & Garcia-Olmedo (eds) 1977), and more recently in legumes (Razdan & Cocking 1981).

As has been discussed by Dewey (1977), wide hybridization can be used in two different ways by the plant breeder. One approach involves the transfer of one of several characters from one species to another (controlled introgression), and the other is the merging of two or more species into a new synthetic species. Both of these approaches are currently being explored by using sexual hybridization: noteworthy successes have resulted in the incorporation of leaf rust resistance of *Aegilops umbellulata* into wheat, *Triticum aestivum* L. (Sears 1956), and the development of the new economically important species triticale, an amphiploid hybrid between *Triticum* and *Secale*. Perhaps it is salutary to note that hybrids between wheat and rye were known as early as 1875; success has not come easily as evidenced by over 40 years' effort by breeders in several countries. What opportunities do protoplasts provide for wide hybridizations such as these? The answer is an opportunity to experiment with the fusion of protoplasts to determine the outcome. These attempts should not be restricted to crosses that are currently impossible sexually. Recent work on the sexual hybridization of rice and sorghum (Deming *et al.* 1979) has highlighted the need for parallel efforts by protoplast fusion. The main difficulty in attempting somatic hybridization between rice and sorghum is the fact that there is at present no regeneration of whole plants from protoplasts of either of these species. There is no difficulty in fusion and limited culture of heterokaryons, but until the present culture difficulties have been overcome we cannot undertake realistic attempts at somatic hybridization. Recent work on the embryo culture of sexual hybrids between maize and sorghum at Cimmyt (James 1980) has also re-emphasized the need for such somatic crosses. Yet again our work in this direction is being impeded by an inability to regenerate whole plants from maize or sorghum protoplasts. There is no difficulty in fusing these protoplasts and obtaining heterokaryons with inherent viability (Brar *et al.* 1980). However, regardless of the method of hybridization, sexual or somatic, the question still remains as to whether the top productive varieties of the main crops of today can still take advantage of the quantitative diversity present in the natural resources. Have we, as suggested by Röbbelen (1979) passed the 'point of no return' at which further addition of foreign variation does nothing but spoil the achieved high level of performance?

What is of importance for plant breeding is that the cytoplasmic mix obtained from protoplast fusions is novel. It would seem likely, as emphasized by McKenzie (1979), that while it is relatively simple to bring together two plastid genotypes in a common cytoplasm by protoplast fusion, it is difficult to keep them together. The inevitable result will be the eventual somatic segregation of the two types into two 'cytoplasmic subspecies', each with only one of the two plastid genotypes in its tissues. Belliard *et al.* (1978) observed that only one species-type of chloroplast DNA remained in the somatic hybrid between two sexually compatible tobacco species. From the studies of Kumar *et al.* (1981) on the polypeptide profiles of fraction 1 protein isolated from somatic hybrid plants, ensuing from the fusion of mesophyll derived protoplasts of wild-type *Petunia parodii* with albino protoplasts of *P. parviflora*, which contained only the chloroplast-coded large subunit polypeptides of *P. parodii*, it has been suggested that the use of albino mutants for selection may cause unidirectional segregation. Clearly an ability to manipulate cytoplasmic (chloroplast) segregation in this way would be very useful, particularly in sexually incompatible species. Opportunities for increasing cytoplasmic variability by protoplast fusions may be greater with mitochondria than with chloroplasts. Belliard *et al.* (1979)

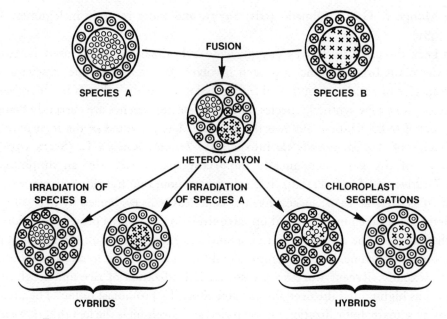

Figure 1. Protoplast fusion and some possible chloroplast segregations.

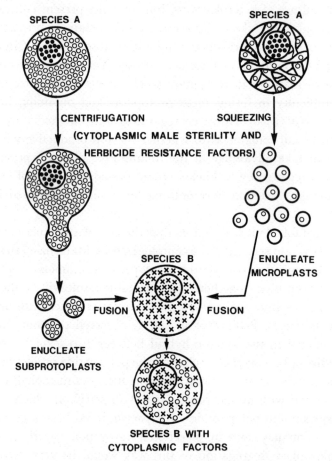

Figure 2. Transfer of cytoplasmic factors by fusion.

obtained evidence for mitochondrial recombination in cytoplasmic hybrids of *N. tabacum* by protoplast fusion. The mitochondrial DNAs of cybrids were different from those of the parent and from the mixture of the two.

If this fusion of protoplasts is also coupled with procedures for the inactivation of the nuclear genome of one of the species, and suitable selection procedures, a range of novel cybrids can theoretically be obtained (figure 1). Aviv *et al.* (1980) have already taken significant steps in this direction in their progeny analysis of the interspecific hybrids *N. tabacum* (cytoplasmic male sterility) and *N. sylvestris* with respect to nuclear and chloroplast markers. Transfer of cytoplasmic male sterility in *Nicotiana* to *N. sylvestris* by protoplast fusion was reported by Zelcer *et al.* (1978), and transfer of cytoplasmic male sterility by protoplast fusion in *Petunia* by Izhar & Power (1979). Rather than irradiating protoplasts to inactivate nuclei, fractionation of protoplasts into enucleate subprotoplasts or enucleate microplasts (Bilkey & Cocking 1980) may provide suitable enucleate units for fusions to produce this novel range of cybrids, thereby avoiding any irradiation effects on the cytoplasm (figure 2). Fractionation of protoplasts may also facilitate the transfer of cytoplasmically based herbicide resistance between sexually incompatible species, as has recently been demonstrated for sexually compatible species with respect to atrazine resistance (Beversdorf *et al.* 1980). The opportunities for the plant breeder are considerable in this respect since cytoplasmic influences are generally quite important in crop yield, for instance in the triticales (Hsam & Larter 1974).

4. Gene transfer

Both sexual and somatic hybridization involve an extensive intermixing of genes, and as emphasized by Riley (1979), there are often disadvantages in attempting to combine the total genetic structures of two species and greater advantage in attempting to incorporate in the recipient crop species some limited, perhaps single, genetic attribute of a donor species. Our earlier work on the fusion of *Petunia* and *Parthenocissus* protoplasts indicated that this might be possible (Power *et al.* 1975).

Currently, advances towards this goal with the use of protoplasts are taking place from several directions. Szabados *et al.* (1980) have recently described the isolation of plant metaphase chromosomes in bulk from highly synchronized cell suspensions and have provided cytological evidence for incorporation of chromosomes into plant protoplasts. Earlier, Malmberg & Griesbach (1980) had described the isolation and purification of chromosomes from plant protoplasts. The demonstrations that, in *Nicotiana*, pollination with highly irradiated (10^5 rad†) killed compatible pollen can cause certain genes from the pollen to be transferred to the egg without proper fertilization has also greatly stimulated work in this direction. This phenomenon, which is in effect transformation of the egg, has resulted in egg transformation for the flower colour gene and the *S* gene (Pandey 1979) and other genes (Virk *et al.* 1977). A particular advantage of this form of sexual transformation is that parthenogenetic plants are produced, avoiding the whole plant regeneration problems currently inherent in the use of protoplasts for many of our crop species. Is the egg special in this respect, or could somatic protoplasts behave similarly? Currently, we are investigating, in collaboration with J. L. Jinks and his colleagues at Birmingham, whether leaf protoplasts of *N. rustica* fused with irradiated

† $1 \text{ rad} = 10^{-2} \text{ Gy} = 10^{-2} \text{ J kg}^{-1}$.

protoplasts will yield plants with similar genetic traits exhibited by those obtained as a result of sexual transformation. We are also currently investigating (in collaboration with A. Müller and his colleagues in Gatersleben, G.D.R., and E. J. Hewitt and his colleagues at Long Ashton) whether such somatic transformation could be used to transfer nitrate reductase genes between different species. We are utilizing the nitrate reductase minus mutants of tobacco isolated by the Gatersleben group and fusing with X-ray irradiated cereal and legume protoplasts, and then selecting for nitrate reductase proficiency among the tobacco cells. Such markers should enable selection of any somatic transformants; in future work, say between a nitrate reductase minus legume or cereal and a wild-type tobacco (or some other species with a highly active nitrate reductase), this selection for limited gene transfer could result in a legume or cereal with improved nitrate reductase capability, and perhaps improved yield. The requirement for auxotrophic mutants for these evaluations emphasizes the need for a greater range of auxotrophic mutants, in a wide range of crop species capable of plant regeneration.

The most direct approach to the introduction of genetic information is afforded by the model of bacterial transformation with purified DNA. This procedure for gene transfer is particularly attractive to the molecular biologist. The opinion of certain plant breeders, however, has been that the utilization of these techniques (even if successful) has definite limitations in scope. For instance, Nelson (1977) has pointed out that these approaches are unlikely to allow selection for traits that arise as a consequence of differentiation or selection in the polygenic complexes that condition yield. It is, however, also recognized that qualitative traits that are expressed in cell cultures enable considerable experimental leverage to be put at the disposal of the investigator. In 1977, summarizing the situation for the prospects for plant genome modification by non-conventional methods, Kleinhofs & Behki concluded that the present science of transformation in plants (including protoplasts) was still vague and poorly defined. They stressed that more and better-controlled experiments were needed to make the results convincing and reproducible in other laboratories, using gene markers that could be specifically identified at the protein or RNA level.

The main deficiencies in previous approaches to plant transformation with purified DNA arose largely because the transforming molecules were not of the type that could be expected to replicate in the plant cell or become integrated into the plant's DNA. Langridge (1978) suggested that a molecule with reasonable chances of acting as a vector would need to be a double-stranded DNA, circular in form, to allow for replication and integration; however, our present knowledge of these requirements is still very limited (Cocking *et al.* 1981). It is well recognized that there are several natural barriers to the introduction, maintenance and expression of 'foreign' DNA in plant cells and that the use of protoplasts greatly facilitates the introduction, but not necessarily the maintenance and expression, of 'foreign' DNA. With cultured animal cells these barriers appear to be less, since it is possible to introduce specific genes into such cultured cells by DNA-mediated gene transfer and to detect the rare transformant by biochemical selection. Such cells can also be co-transformed with two physically unlinked genes. Wigler *et al.* (1979) have shown that co-transformed cultural animal cells can be identified and isolated when one of these genes codes for a selectable marker, such as thymidine kinase; ligation to a viral vector was not required.

The singular attraction of *Agrobacterium tumefaciens* is that when interacting with plants it naturally manages to transfer, maintain and express its prokaryotic DNA in plant cells. Numerous workers are now busy trying to exploit this natural capability of the *Agrobacterium*

bacterium to use the *Agrobacterium tumefaciens* Ti plasmid as a host vector system for introducing foreign DNA in plant cells (Drummond 1979; Hernalsteens *et al.* 1980). In many of these experiments on plant transformation it is not necessary, at least initially, to use protoplasts.

Aiding this work has been the demonstration that plasmids isolated from *Agrobacterium* can be used to transform protoplasts, transformed cells being selected by their independence and associated proliferation on hormone-free media, overgrowth formation when grafted onto host plants, octopine synthesis and lysopine dehydrogenase activity (Davey *et al.* (1980a), and that intact *Agrobacterium* will transform cells that have been formed by the regeneration of a wall on isolated protoplasts of wild-type *N. tabacum* (Davey *et al.* 1980b) (figure 3).

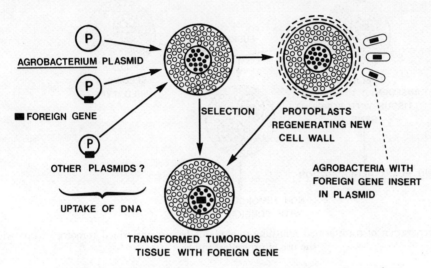

FIGURE 3. Strategy for transformation with the use of protoplasts.

Opportunities for the use of protoplasts will arise in this work when an inadequate percentage of the cells are transformed in the tumours from plant–*Agrobacterium* transformations. This will necessitate cloning procedures, either from the transformants, or by initially using protoplast-regenerant cells interacting with intact agrobacteria (Davey *et al.* 1980b). If, with further work, it is established that foreign gene transfer and expression into dicotyledonous plants is possible with such an *Agrobacterium* vector system, the need to extend this approach to the cereals will be paramount. It would seem probable from the work of Davey *et al.* (1980a, b) that cereal protoplasts interacted with isolated *Agrobacterium* plasmids will be capable of being transformed, since the removal of the cell wall is likely to remove the natural barrier to infection. Protoplast-regenerant cells from cereals are also likely to be transformed by intact agrobacteria since the newly synthesized cell wall formed on protoplasts is likely to facilitate bacterial interaction.

It has often been suggested that it is of little practical value to the plant breeder to engineer plant tumours, however enriched they might be with foreign genes. Regeneration of plants is an essential prerequisite for uses in plant breeding. Protoplasts will undoubtedly play a key role in resolving this type of difficulty. It has been shown in fusion experiments between protoplasts from wild type and tumour cell lines (which frequently fail to regenerate into plants) that fusion between the morphogenic and non-morphogenic cells results in shoot-forming callus – the ability to regenerate shoots is dominant when one of the partners is a tumour cell. It would seem probable that the fusion of protoplasts, subprotoplasts or microplasts (Bilkey & Cocking

1980) from undifferentiated tumour tissue carrying foreign genes, and of protoplasts prepared from wild-type plants, will result in regenerated normal plants in which foreign genes are expressed (Wullems *et al.* 1980) (figure 4).

An *Escherichia coli* plasmid vehicle would not be capable of interacting with walled cells, and plant protoplasts will be essential for any attempts at transformation with, for instance, dominant drug resistance genes from bacteria, if no *Agrobacterium* vector is used. With cereal protoplasts it may be essential to use a plasmid vehicle other than that of *Agrobacterium* (such as that of *E. coli*), if natural transformation of cereal protoplasts by *Agrobacterium* or its plasmid should prove impossible.

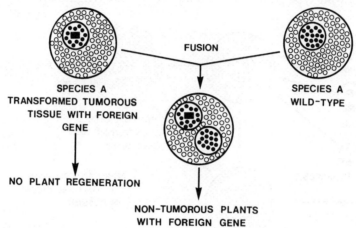

FIGURE 4. Regeneration of transformed non-tumorous plants from transformed tumorous tissue with the use of protoplast fusion.

5. CONCLUSIONS

From what has already been described about the opportunities from the use of protoplasts for plant breeding, it will be evident that much more work will be required before the use of protoplasts adds significantly to the armoury of the plant breeder. What is clearly required is a continuing dialogue with plant breeders to ensure that the objectives of manipulations with protoplasts are closely aligned to those of the breeder. The contribution from the use of protoplasts will only be significant if the existing breeding problems and objectives with the use of conventional methods are uppermost in the thinking and manipulations of protoplast workers.

REFERENCES (Cocking)

Aviv, D., Fluhr, R., Edelman, M. & Galun, E. 1980 Progeny analysis of the interspecific hybrids: *Nicotiana tabacum* (CMS) + *Nicotiana sylvestris* with respect to nuclear and chloroplast markers. *Theor. appl. Genet.* **56**, 145–150.

Belliard, G., Pelletier, G., Vedel, F. & Quetier, F. 1978 Morphological characteristics and chloroplast DNA distribution in different cytoplasmic parasexual hybrids in *Nicotiana tabacum*. *Molec. gen. Genet.* **165**, 231–237.

Belliard, G., Vedel, F. & Pelletier, G. 1979 Mitochondrial recombination in cytoplasmic hybrids of *Nicotiana tabacum* by protoplast fusion. *Nature, Lond.* **281**, 401–403.

Beversdorf, W. D., Weiss-Lerman, J., Erickson, L. R. & Souza Machado, V. 1980 Transfer of cytoplasmically-inherited triazine resistance from bird's rape to cultivated oilseed rape (*Brassica campestris* and *B. napus*). *Can. J. Genet. Cytol.* **22**, 167–172.

Bhojwani, S. S., Evans, P. K. & Cocking, E. C. 1977 Protoplast technology in relation to crop plants: progress and problems. *Euphytica* **26**, 343–360.

Bilkey, P. C. & Cocking, E. C. 1980 Isolation and properties of plant microplasts: newly identified subcellular units capable of wall synthesis and division into separate micro cells. *Eur. J. Cell Biol.* **22**, 502.

Brar, D. S., Rambold, S., Constabel, F. & Gamborg, O. L. 1980 Isolation, fusion and culture of sorghum and corn protoplasts. *Z. PflPhysiol.* **96**, 269–275.

Cocking, E. C. 1979 Parasexual reproduction in flowering plants. *N.Z. Jl Bot.* **17**, 665–671.

Cocking, E. C., Davey, M. R., Pental, D. & Power, J. B. 1981 Plant genetic manipulation. *Nature, Lond.* (In the press.)

Cocking, E. C. & Riley, R. 1981 Application of tissue culture and somatic hybridisation to plant improvement. In *Plant Breeding Symposium II*, Iowa State University, March 1979 (ed. K. J. Frey). (In the press.)

Davey, M. R., Cocking, E. C., Freeman, J., Pearce, N. & Tudor, I. 1980a Transformation of *Petunia* protoplasts by isolated *Agrobacterium* plasmids. *Pl. Sci. Lett.* **18**, 307–413.

Davey, M. R., Cocking, E. C., Freeman, J., Draper, J., Pearce, N., Tudor, I., Hernalsteens, J. P., De Beuckeleer, M., Van Montagu, M. & Schell, J. 1980b The use of plant protoplasts for transformation by *Agrobacterium* and isolated plasmids. In *Advances in Protoplast Research* (Proceedings of the 5th Int. Protoplast Symposium, Szeged, Hungary, 1979), pp. 425–430. Budapest: Akadémiai Kiadó.

Deming, Z., Lanfang, D., Shanbao, C., Xianbin, S. & Xialon, D. 1979 Diversity and specific performance of progenies from distant hybridization between rice and sorghum. *Acta genet. sinica* **6**, 419–420.

Dewey, D. R. 1977 The role of wide hybridization in plant improvement. In *Genetics lectures* (ed. R. Bogart), vol. 5, pp. 7–18. Oregon State University Press.

Drummond, M. 1979 Crown gall disease. *Nature, Lond.* **281**, 343–347.

Evans, P. K. & Cocking, E. C. 1974 The techniques of plant cell culture and somatic hybridization. In *New techniques in biophysics and cell biology* (ed. R. H. Pain & B. J. Smith), vol. 2, pp. 127–158. London: John Wiley.

Gleba, Y. Y. & Hoffmann, F. 1979 'Arabidobrassica' plant genome engineering by protoplast fusion. *Naturwissenschaften* **66**, 547–554.

Hernalsteens, J. P., Van Vliet, F., De Beuckeleer, M., Depicker, A., Engler, G., Lemmers, M., Holsters, M., Van Montagu, M. & Schell, J. 1980 The *Agrobacterium tumefaciens* Ti plasmid as a host vector system for introducing foreign DNA in plant cells. *Nature, Lond.* **287**, 654–656.

Hsam, S. L. K. & Larter, E. N. 1974 Influence of source of wheat cytoplasm on the nature of proteins in hexaploid triticale. *Can. J. Genet. Cytol.* **22**, 167–172.

Izhar, S. & Power, J. B. 1979 Somatic hybridisation in *Petunia*: a male sterile cytoplasmic hybrid. *Pl. Sci. Lett.* **14**, 49–55.

James, J. 1980 New hybrids between maize and sorghum. *Euphytica*. (In the press.)

Kleinhofs, A. & Behki, R. 1977 Prospects for plant genome modification by nonconventional methods. *A. Rev. Genet.* **11**, 79–101.

Krumbiegel, G. & Schieder, O. 1979 Selection of somatic hybrids after fusion of protoplasts from *Datura innoxia* Mill and *Atropa belladona* L. *Planta* **145**, 371–379.

Kumar, A., Wilson, D. & Cocking, E. C. 1981 Polypeptide composition of Fraction 1 protein of the somatic hybrid between *Petunia parodii* and *Petunia parviflora*. *Biochem. Genet.* **19**, 255–261.

Langridge, J. 1978 Molecular vectors for plant cells. In *Proceedings of Symposium on Plant Tissue Culture*, pp. 391–396. Peking: Science Press. (Guozi Shudian, P.O. Box 399, Peking, China.)

McKenzie, R. J. 1979 Organelle genetics. *BioScience* **29**, 569–572.

Malmberg, R. & Griesbach, R. J. 1980 The isolation of mitotic and meiotic chromosomes from plant protoplasts. *Pl. Sci. Lett.* **17**, 141–147.

Melchers, G., Sacristan, M. D. & Holder, A. A. 1978 Somatic hybrid plants of potato and tomato regenerated from fused protoplasts. *Carlsberg Res. Commun.* **43**, 203–218.

Nelson, O. E. 1977 The applicability of plant cell and tissue culture techniques to plant improvement. In *Molecular genetic modification of eukaryotes* (ed. I. Rubenstein, R. L. Phillips, C. E. Green & R. J. Desnick), pp. 67–76. San Francisco: Academic Press.

Pandey, K. K. 1979 Gametic gene transfer in *Nicotiana* by means of irradiated pollen. *Genetica* **49**, 53–69.

Power, J. B., Frearson, E. M., Hayward, C. & Cocking, E. C. 1975 Some consequences of the fusion and selective culture of *Petunia* and *Parthenocissus* protoplasts. *Pl. Sci. Lett.* **5**, 197–207.

Power, J. B. & Davey, M. R. 1979 *Laboratory manual: plant protoplasts (isolation, fusion, culture, genetic transformation)*. Department of Botany, University of Nottingham, University Park, Nottingham NG7 2RD, U.K.

Power, J. B., Berry, S. F., Chapman, J. V. & Cocking, E. C. 1980 Somatic hybridization of sexually incompatible *Petunias*: *Petunia parodii*, *Petunia parviflora*. *Theor. appl. Genet.* **57**, 1–4.

Razdan, M. K. & Cocking, E. C. 1981 Genetic improvement of legumes by exploring extra-specific genetic variation. *Euphytica*. (In the press.)

Riley, R. 1979 Methods for the future. In *Plant breeding perspectives* (ed. J. Sneep & A. J. T. Hendriksen), pp. 321–369. Wageningen: Centre for Agricultural Publication and Documentation.

Röbbelen, G. 1979 Transfer of quantitative characters from wild and primitive forms. In *Proc. Conf. on Broadening Genetic Base of Crops* (ed. A. C. Zeven & A. M. Van Harten), pp. 249–255. Wageningen: Centre for Agricultural Publication and Documentation.

Sanchez-Monge, E. & Garcia-Olmedo, F. (eds) 1977 *Interspecific hybridization in plant breeding* (Proceedings of the 8th Congress of Eucarpia). Madrid: Escuela Tecnica Superior de Ingenieros Agronomos Universidad Politecnica de Madrid.

Santos, A. V. P. dos, Outka, D. E., Cocking, E. C. & Davey, M. R. 1980 Organogenesis and somatic embryogenesis in tissues derived from leaf protoplasts and leaf explants of *Medicago sativa*. *Z. PflPhysiol.* **99**, 261–270.

Scowcroft, W. & Larkin, P. J. 1980 Tissue culture research: current status and future potential. In *Special Planning Conference on Rice Tissue Culture*, pp. 1–12. Los Banos Laguna, Philippines: The International Rice Research Institute.

Sears, E. R. 1956 The transfer of leaf-rust resistance from *Aegilops umbellulata* to wheat. *Brookhaven Symp. Biol.* **9**, 1–22.

Shepard, J. F., Bidney, D. & Shahin, E. 1980 Potato protoplasts in crop improvement. *Science, N.Y.* **208**, 17–24.

Szabados, L., Hadlaczky, G. & Dudits, D. 1980 Uptake of isolated plant chromosomes by plant protoplasts. *Expl Cell Res.* (In the press.)

Virk, D. S., Dhahi, S. J. & Brumpton, R. J. 1977 Matromorphy in *Nicotiana rustica*. *Heredity, Lond.* **39**, 287–295.

Wenzel, G., Schieder, O., Przewozny, T., Sopory, S. K. & Melchers, G. 1979 Comparison of single cell culture derived *Solanum tuberosum* L. plants and a model for their application in breeding programmes. *Theor. appl. Genet.* **55**, 49–55.

Wernicke, W. & Brettell, R. 1980 Somatic embryogenesis from *Sorghum bicolor* leaves. *Nature, Lond.* **287**, 138–139.

Wigler, M., Sweet, R., Sim, G. K., Wold, B., Pellicer, A., Lacy, E., Maniatis, T., Silverstein, S. & Axel, R. 1979 Transformation of mammalian cells with genes from procaryotes and eucaryotes. *Cell* **16**, 777–785.

Wullems, G. J., Molendijk, L. & Schilperoort, R. A. 1980 The expression of tumour markers in intraspecific somatic hybrids of normal and crown gall cells from *Nicotiana tabacum*. *Theor. appl. Genet.* **56**, 203–208.

Zelcer, A., Aviv, D. & Galun, E. 1978 Interspecific transfer of cytoplasmic male sterility by fusion between protoplasts of normal *Nicotiana sylvestris* and X-ray irradiated protoplasts of male sterile *N. tabacum*. *Z. PflPhysiol.* **90**, 397–407.

Discussion

E. Thomas (*Rothamsted Experimental Station, Harpenden, Herts., U.K.*). Professor Cocking stated that a special feature of plant protoplast culture was the variation in agronomic characters observable in regenerated plants, which is undetectable by other tissue culture systems; potato was given as the example. Recent, but as yet unpublished, data from Holland, very clearly demonstrates in potato that a large amount of variation also occurs from calluses of complex explant origin. Thousands of variant plants have been obtained from such calluses, probably indicating cytological instability during callus culture rather than a unique feature of plant protoplasts. Because of the comparatively difficult nature of protoplast technique in crop plants, perhaps it would be easier, cheaper and far more rapid to obtain variant plants by more conventional tissue culture methods. Further, the use of such variants in plant breeding may be questionable, if we cannot direct a selection pressure for the particular agronomic character that we are seeking without adversely affecting other important agronomic characters. This may be particularly true of vegetatively propagated crops such as potato.

E. C. Cocking. This large amount of variation occurring from calluses of complex explant origin is not unexpected, and re-emphasizes that plant tissue culture *per se* appears to be an unexpectedly new and novel source of genetic variation. Only further detailed studies on a range of potato varieties will reveal whether *extra* variation is revealed as a result of protoplast cloning. Generally, in plant breeding it is difficult to direct a selection pressure for a particular agronomic character without adversely affecting other important agronomic characters; those studying variation in plants regenerated from potato protoplasts have themselves highlighted this type of difficulty. These questions will only be resolved after suitable field trials in which such plants are exposed to environmental pressures to eliminate any that are simultaneously deficient in some critical feature.

Chromosomal DNA in higher plants

By H. Rees, F.R.S., and R. K. J. Narayan

Department of Agricultural Botany, University College of Wales, Penglais, Aberystwyth SY23 3DD, U.K.

There is an astonishing variation in the amount of chromosomal DNA among species of higher plants. Much of this variation is due to the amplification of base sequences within the chromosomes. As a result, the amount of DNA in the nuclei of many species is very great. In particular, the chromosomes are rich in repetitive DNA, which may comprise 70% or more of the total. This fraction contains at most only a few genes that code for proteins. What, then, is its functional significance? There is evidence that DNA amount *per se* affects cell size and the duration of cell divisions. The results of recent assays also provide evidence that particular repetitive sequences have specific effects upon the phenotype. While these effects may be small in themselves they may nevertheless be important in Nature and in plant breeding.

1. Introduction

Among species of higher plants there is an astonishing variation in nuclear DNA amount. Much of it is due to polyploidy, but there is a widespread and substantial variation resulting from the amplification and, perhaps, deletion of DNA segments within the chromosomes. The extent of such variation is revealed by the survey of Bennett & Smith (1976). Among diploid flowering plants we have, at one extreme, species like *Lotus corniculatus* with less than 1 pg of DNA in $2C$ nuclei. At the other extreme there are species of the genus *Fritillaria* with nuclear DNA amounts of 180 pg. The range is 300-fold. Investigations of species within a number of genera and families of flowering plants provide some of the answers that bear upon the following problems.

(1) What is the structural basis of the DNA variation within chromosomes?

(2) What functional significance, if any, has the DNA that results from the amplification of base sequences that often accompanies the divergence and evolution of species?

We have answers to account for many aspects of the structural changes associated with the variation in chromosomal DNA. The question of function with respect to the DNA variation remains largely unresolved. This is true not only for higher plants but for eukaryotes in general. In short, the C-value paradox has yet to be explained. The question of function is of concern not only from the standpoint of natural selection and adaptation. Now that methods have been developed by which DNA fragments may be excised and cloned in alien species it is imperative to find out which of the chromosomal DNA fractions in eukaryotes may be of use to the applied scientist, the plant breeder and others, for the improvement of plant and animal products.

2. DNA variation in *Lathyrus*

(a) Quantitative change

The results of our own investigations among species within the genus *Lathyrus* display many of the features characteristic of the quantitative nuclear DNA variation associated with the divergence and evolution of flowering plant species. Table 1 and figure 1a show the distribution of nuclear DNA amounts in *Lathyrus* species. All are diploids with 14 chromosomes. Two observations may be made.

(a) Despite their close taxonomic affinity there is, roughly, a threefold variation among species. It is tempting, on the face of it, to argue that much of the chromosomal DNA must be redundant or inert. The more specialized inbreeding annual species, have, on average, a significantly lower DNA content than the outbreeders. Since inbreeders must be derived from outbreeding ancestors we have to conclude that the evolution of *Lathyrus* species may have been accompanied by a massive DNA diminution (Rees & Hazarika 1969).

(b) There is an element of discontinuity associated with the DNA distribution. The discontinuity is, moreover, of a surprisingly consistent pattern. There is an interval of approximately 3 pg between each group. The variation, in other words, is achieved by a series of

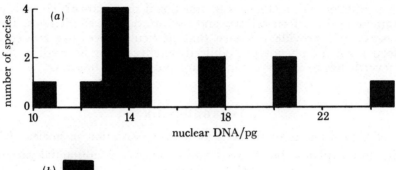

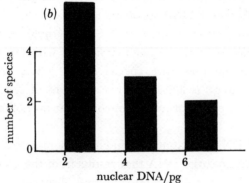

FIGURE 1. The distribution of nuclear DNA amounts among (a) *Lathyrus* species and (b) species of the genus *Clarkia*.

TABLE 1. NUCLEAR DNA AMOUNTS (PICOGRAMS) IN 2C NUCLEI OF *LATHYRUS* SPECIES

	total DNA	DNA in heterochromatin	DNA in euchromatin	non-repetitive DNA	repetitive DNA
L. angulatus	10.90	1.60	9.30	4.36	6.54
L. articulatus	12.45	1.90	10.55	5.48	6.97
L. nissolia	13.20	2.77	10.43	5.41	7.79
L. clymenum	13.75	2.91	10.84	5.23	8.52
L. ochrus	13.95	3.48	10.47	5.58	8.37
L. aphaca	13.97	—	—	5.17	8.90
L. cicera	14.18	3.36	10.82	5.96	8.22
L. sphaericus	14.18	—	—	—	—
L. sativus	17.15	4.50	12.65	5.83	11.32
L. odoratus	17.16	—	—	5.67	11.49
L. hirsutus	20.27	7.49	12.78	6.17	14.20
L. tingitanus	22.18	7.39	14.79	7.90	14.28
L. sylvestris	24.26	8.70	15.56	7.27	16.99

'quantum jumps'. It appears as if changes involving only particular DNA packages were tolerable during speciation. This could be interpreted as an adaptive constraint upon the quantitative DNA variation. If so, the variation must be of functional significance. Similar 'quantum jumps' have been reported by Rothfels *et al.* (1966) in the Anemonae and by Martin & Shanks (1966) in *Vicia*. We have recently found a particularly striking example in the genus *Clarkia* (figure 1b).

(b) *The distribution of quantitative DNA change*

Comparisons between the chromosomes of species with low and high nuclear DNA amounts indicate that the quantitative DNA changes implicate all chromosomes within the complement (Rees & Hazarika 1969). The same is true for other genera, e.g. *Allium* (Jones & Rees 1968) and *Lolium* (Rees & Jones 1967). However, different chromosome components are affected to different degrees.

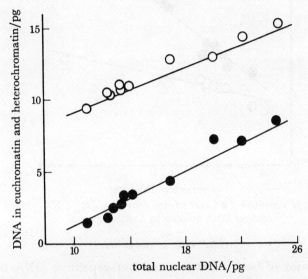

FIGURE 2. The amount of DNA located in euchromatin (○) and in heterochromatin (●) plotted against the total nuclear DNA amount in *Lathyrus* species.

(c) *Heterochromatin and euchromatin*

Table 1 and figure 2 show the amounts of DNA located in euchromatin and in heterochromatin in 10 *Lathyrus* species differing in total nuclear DNA amount. With increasing total DNA it will be seen that, while there is an increase in both euchromatin and heterochromatin, the rate of increase is greater within heterochromatin than within euchromatin. We emphasize, in particular, the linearity of the regressions in figure 2. It signifies that the extra, supplementary DNA is of a consistent composition throughout the genus. The addition of each picogram of DNA located within euchromatin is accompanied by an additional 1.2 pg located within heterochromatin. This constancy, we again suggest, implies a rigid constraint upon the kind of DNA variation tolerable during divergence and speciation (Rees & Narayan 1977).

(d) *Repetitive and non-repetitive DNA*

Table 1 and figure 3 show the amounts of non-repetitive and repetitive DNA in *Lathyrus* species with different DNA amounts. An increase in the total DNA amount is due to an increase in both non-repetitive and repetitive components, but preponderantly the latter. We emphasize again the linearity of the regression lines, showing that the composition of the supplementary DNA is constant throughout the genus. For every picogram of non-repetitive DNA there is an increase of 4.2 pg of repetitive DNA. The constancy, like that which applies to DNA in euchromatin relative to heterochromatin, suggests yet again a rigid constraint upon the nature of the DNA changes associated with divergence and evolution. The constraints are at least an indication that the variation may be of adaptive and functional significance (Hutchinson *et al.* 1980).

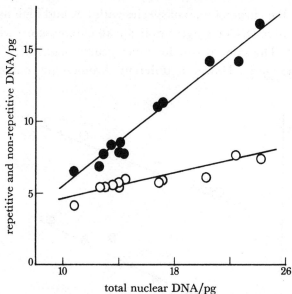

FIGURE 3. The amount of repetitive (●) and of non-repetitive (○) DNA plotted against the total nuclear DNA amount in *Lathyrus* species.

(e) Base sequence composition

Variation in the amount of both repetitive and non-repetitive DNA is accompanied by changes in base sequence composition (Narayan & Rees 1977), for example a heavy satellite is readily distinguished when the DNA of *L. tingitanus* is bound to actinomycin D and centrifuged in a caesium chloride gradient (figure 4). *In situ* DNA–DNA hybridization shows that the repeated sequence that makes up this satellite is located in heterochromatic blocks adjacent to the centromeres of *L. tingitanus* chromosomes. The satellite DNA does not 'hybridize' to the DNA in chromosomes of other species, e.g. *L. hirsutus*. The base sequence is either rare or absent from this species.

Detailed information about the composition, variation and the distribution of satellite DNA in flowering plants, particularly cereals, are given by Flavell *et al.* (1980) and Gerlach & Peacock (1980). For the present it will suffice to emphasize that a change in the amount of DNA during speciation is accompanied by a change in base composition.

We would expect, of course, to find some base sequences to be conserved, particularly those associated with structural genes. We need to bear in mind, however, that the amount of chromosomal DNA that contributes to the structural genes is small. Estimates of the number of structural genes in the chromosomes of eukaryotes range from 10 000 to 30 000 (see Cavalier-Smith 1978), which would implicate less than 5% of the DNA in *Lathyrus* and many other species of flowering plants. It is the function of the remaining 95% or more of the chromosomal DNA, composed of both repetitive and non-repetitive base sequences, that remains an enigma. We shall now consider the results of some experiments that bear upon this problem.

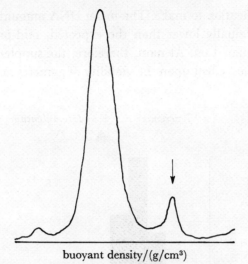

buoyant density/(g/cm³)

FIGURE 4. A 'heavy' DNA satellite (arrowed) in *Lathyrus tingitanus* revealed after binding with actinomycin D and ultracentrifugation in a CsCl gradient.

3. The consequences of variation in DNA amount

(a) Cell size and cell division

There is a large body of evidence showing that the duration of mitotic cycles in root meristems of flowering plants increases with increasing nuclear DNA amount (Van't Hof 1965; Evans *et al.* 1972). In root meristems, also, the cell size and mass increase with increasing DNA amount (Martin 1966). The dependence of these important components of growth on the DNA amount might well be expected to have important adaptive implications. In this connection, Bennett (1972) has pointed out that short-lived, ephemeral plants have, in general, lower nuclear DNA amounts than long-lived perennials. There is, however, a puzzling element in this relation. At 20 °C a change of 1 h in the duration of mitotic cycles in root meristems of dicotyledons is achieved by loss or gain of 3 pg of nuclear DNA (Evans *et al.* 1972). Are we, in the light of this fact, to conclude that control over the duration of mitotic cycles and, for that matter, of cell size is achieved by variation in nuclear DNA? If so, the control invokes massive alterations in chromosome material; on the face of it a crude and, materially, expensive strategy. Even so, a powerful argument to this effect has, indeed, been presented by Cavalier-Smith (1978).

(b) An assay of DNA effects in Lolium

Leaving aside the consequences of DNA variation upon cell cycles and cell size, what other effects can we attribute to the DNA that is amplified, or deleted, in conjunction with speciation in flowering plants? The nuclei of inbreeding species of *Lolium* have 40 % more DNA than those of outbreeders. All are diploids with 14 chromosomes. Despite this large DNA difference, the species produce viable F_1 hybrids, and fertile hybrids at that.

Viability

Figure 5 shows estimates of DNA amounts in *L. perenne* (an outbreeder, *L. temulentum* (an inbreeder), their F_1 hybrid and their backcross progenies. The DNA values in the backcross follow a normal distribution, ranging from that of the 'low' DNA parent to the mid-parent value. We conclude that the segregation of the supplementary DNA fraction, that which distinguishes *L. perenne* from *L. temulentum*, has no effect upon the viability of gametes or of zygotes.

There is one slight qualification to make. The mean DNA amount for the progeny of this and other backcrosses is marginally lower than the expected, mid-parent value. The departure from expectation is less than 1 pg. At most, therefore, the supplementary DNA derived from *L. temulentum* has a marginal effect upon the viability of gametes and zygotes (Hutchinson *et al.* 1979).

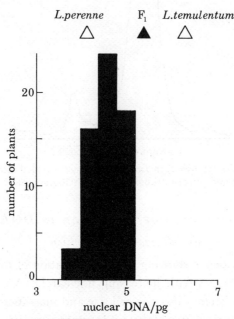

FIGURE 5. The distribution of nuclear DNA amounts in *Lolium perenne*, *L. temulentum*, their F_1 hybrid and the backcross progeny of the F_1 to *L. perenne*.

Chiasma frequencies in pollen mother cells

The extra, supplementary DNA in inbreeding *Lolium* species results from amplification of the DNA within each chromosome of the complement. This is made clear by the structure of bivalents at first metaphase of meiosis in the F_1 hybrid. All seven bivalents are asymmetrical (figure 6). The question is: how does the supplementary DNA affect homology, the capacity to pair effectively and form chiasmata at meiosis in hybrids and hybrid derivatives? Two kinds of effects on chromosome pairing and chiasma formation might be envisaged. First, there is a gross structural dissimilarity between the parental chromosomes due to the extra DNA in *L. temulentum* chromosomes. Secondly, there is the possibility that the extra DNA might carry determinants that influence chiasma formation. The mean chiasma frequencies for *L. perenne* (the outbreeder), *L. temulentum* (the inbreeder) and their F_1 hybrid are 11.6, 13.9 and 11.0 respectively (see also figure 7). The mean chiasma frequency of the F_1 hybrid is not significantly different from that of *L. perenne*, despite the structural dissimilarity of the parental chromosomes. Figure 7 shows the chiasma frequencies in plants of the progeny of the backcross, $F_1 \times L.$ *perenne*, plotted against the nuclear DNA amount. There is a wide range of chiasma frequencies. This is to be expected as the result of the segregation of parental genes controlling chiasma frequency. There is, however, no correlation between the chiasma frequencies and the DNA amounts. There is therefore no evidence that the chiasma frequency is influenced by determinants located in the supplementary DNA, nor is there any evidence that the chiasma frequency is influenced by the gross structural differences within bivalents that result from the supplementary DNA deriving from the *L. temulentum* parent.

The results of DNA assays of the kind that we have described, involving many phenotypic characters in backcrosses and F_2 progenies from interspecific *Lolium* hybrids, were published by Hutchinson *et al.* (1980). The characters ranged from seedling height to flowering time. Apart from a very slight effect upon gamete or zygote viability, the assays showed that the large

FIGURE 6. Asymmetrical bivalents at first metaphase in the hybrid *Lolium temulentum* × *L. perenne*.

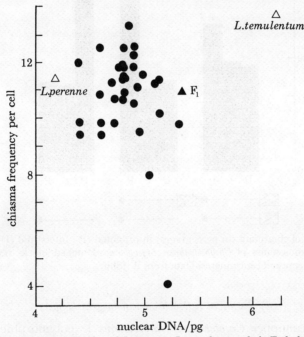

FIGURE 7. The mean chiasma frequencies of *L. perenne*, *L. temulentum*, their F_1 hybrid and the backcross progeny, plotted against the nuclear DNA amount. Data from A. Seal.

amount of supplementary DNA that distinguishes the inbreeding from outbreeding species has surprisingly little effect upon growth and development. Does this mean that much of this supplementary DNA is genetically inert, so-called selfish or parasitic DNA (Doolittle & Sapienza 1980; Orgel & Crick 1980)? Not necessarily. There are certain limitations to the assays in *Lolium*, at least for detecting relatively small effects, and small effects may be important. For detecting such effects we suggest that it would be more effective to identify and assay particular DNA base sequences rather than to assay massive amounts of chromosomal DNA as in the *Lolium* investigation.

(c) Knobbed 10 in maize

A small heterochromatic knob in the long arm distinguishes the abnormal from the normal form of chromosome 10, the smallest chromosome in the maize complement. The presence of the knob on 10 has a dramatic effect on chromosome behaviour at both divisions of meiosis. The ends of the chromosomes, particularly of the long arms, display 'centromere-like' activity, i.e. they develop spindle fibres and move towards the spindle poles (Rhoades 1952). The chromosomes will respond to the presence of abnormal 10, however, only if they themselves bear heterochromatin of a particular DNA composition, namely a highly repetitive sequence of 185 base pairs (Peacock 1981). Classes of heterochromatin lacking this 185 base-pair repeat do not respond.

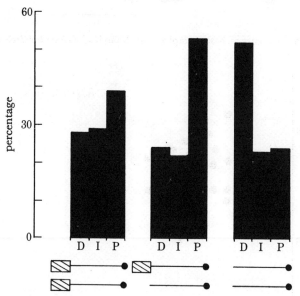

FIGURE 8. The distribution of chiasmata (in percentages) in proximal (P), interstitial (I) and distal (D) segments of medium-length chromosomes of *Cryptobothrus chrysophorus*. Hatched blocks represent heterochromatic segments, black discs represent centromeres. Data from B. John.

(d) Terminal heterochromatin in grasshoppers

The Australian grasshopper *Cryptobothrus chrysophorus* is polymorphic for large terminal blocks of heterochromatin on three of the large and six of the medium-sized chromosomes of the complement. The blocks are distal, i.e. at the chromosome ends furthest from the centromeres (figure 8). In the absence of blocks the chiasmata are located mainly in the distal euchromatin. In bivalents homozygous or heterozygous for the heterochromatic blocks, there is a dramatic shift in chiasma distribution, away from the blocks, proximally towards the centromeres (John & King 1980). It is possible that the heterochromatin interferes directly with the effective pairing of chromosomes at pachytene. If so, the repetitive DNA of which it is composed achieves its effect upon the phenotype not through transcription and translation of a product but by mechanical modifications of the chromosome architecture. The same may well be true of heterochromatic knobs in maize (Rhoades 1952). The maize story tells us, however, that such mechanical modifications may nevertheless depend on specific DNA sequences.

4. Discussion

The results of Peacock (1981) and of John & King (1980) in §§3c and 3d are encouraging indications that assays of particular chromosome DNA sequences may provide useful information about the functional properties of that large, mainly repetitive DNA fraction of eukaryote chromosomes that, on the face of it, appears to be relatively inert. This is not to say, of course, that all of it is active. Some of it may be, as many have asserted, quite inert and 'selfish'. After all the only requirement for the spread and establishment of a selfish DNA segment is that it multiplies at a faster rate than the normal. On this score a brief reference to supernumerary, B chromosomes, which are common in higher plants, is both apposite and cautionary. B chromosomes, even when deleterious to the fitness of the organism, may spread and persist in a population because they have the capacity of increasing their number in gametic nuclei in comparison with 'ordinary' chromosomes of the complement. In this respect their DNA could qualify for the category of selfish or parasitic (Ostergren 1945). Yet this does not mean that all of the B chromosome DNA is inactive: there is ample evidence to the contrary (Jones 1975). Under certain circumstances, indeed, there is good evidence that it contributes to an increase in fitness (Rees & Hutchinson 1973). We would argue that the same may be true of much of the repetitive and other fractions of the chromosomal DNA of eukaryotes that present us with the riveting challenge of the C-value paradox. The argument will only gather substance, however, when we have the results of more genetic assays to analyse and evaluate.

We are grateful to Mr Alan Seal and to Professor B. John for permission to quote their experimental results.

References (Rees & Narayan)

Bennett, M. D. 1972 Nuclear DNA amount and minimum generation time in herbaceous plants. *Proc. R. Soc. Lond.* B **191**, 109–135.

Bennett, M. D. & Smith, J. B. 1976 Nuclear DNA amounts in angiosperms. *Phil. Trans. R. Soc. Lond.* B **274**, 227–274.

Cavalier-Smith, T. 1978 Nuclear volume control by neucleoskeletal DNA, selection for cell volume and cell growth rate, and the solution of the DNA C-value paradox. *J. Cell Sci.* **34**, 247–278.

Doolittle, W. F. & Sapienza, C. 1980 Selfish genes, the phenotype paradigm and genome evolution. *Nature, Lond.* **284**, 601–603.

Evans, G. M., Rees, H., Snell, C. L. & Sun, S. 1972 The relationship between nuclear DNA amount and the duration of the mitotic cycle. In *Chromosomes today*, vol. 3, pp. 24–31. London: Longman.

Flavell, R. B., Bedbrook, J. R., Gerlach, W. G., Jones, J., Dyer, T. A., O'Dell, M. & Thompson, R. D. 1980 Molecular events in the evolution of cereal chromosomes. In *The plant genome* (4th John Innes Symposium), pp. 15–30. John Innes Charity.

Gerlach, W. G. & Peacock, W. J. 1980 Chromosomal locations of highly repeated DNA sequences in wheat. *Heredity, Lond.* **44**, 269–276.

Hutchinson, J., Rees, H. & Seal, A. G. 1979 An assay of the activity of supplementary DNA in *Lolium*. *Heredity, Lond.* **43**, 411–421.

Hutchinson, J., Narayan, R. K. J. & Rees, H. 1980 Constraints upon the composition of supplementary DNA. *Chromosoma* **78**, 137–145.

John, B. & King, M. 1980 Heterochromatin variation in *Cryptobothrus chrysophorus*. III. Synthetic hybrids. *Chromosoma* **78**, 165–186.

Jones, R. N. 1975 B-chromosome systems in flowering plants and animal species. *Int. Rev. Cytol.* **40**, 1–100.

Jones, R. N. & Rees, H. 1968 Nuclear DNA variation in *Allium*. *Heredity, Lond.* **23**, 591–605.

Martin, P. G. 1966 Variation in the amounts of nucleic acids in the cells of different species. *Expl Cell Res.* **44**, 84–90.

Martin, P. G. & Shanks, R. 1966 Does *Vicia faba* have multistranded chromosomes? *Nature, Lond.* **211**, 650–651.

Narayan, R. K. J. & Rees, H. 1977 Nuclear DNA divergence among *Lathyrus* species. *Chromosoma* **63**, 101–107.

Orgel, L. E. & Crick, F. H. C. 1980 Selfish DNA: the ultimate parasite. *Nature, Lond.* **284**, 604–607.

Ostergren, G. 1945 Parasitic nature of extra fragment chromosomes. *Bot. Notiser* **2**, 157–163.
Peacock, W. J. 1981 In *Chromosomes today*, vol. 7. (In the press.)
Rees, H. & Hazarika, M. H. 1969 Chromosome evolution in *Lathyrus*. In *Chromosomes today*, vol. 2 (ed. K. R. Lewis & C. D. Darlington), pp. 158–165. Edinburgh: Oliver & Boyd.
Rees, H. & Hutchinson, J. 1973 Nuclear DNA variation due to B chromosomes. *Cold Spring Harb. Symp. quant. Biol.* **38**, 175–182.
Rees, H. & Jones, G. H. 1967 Chromosome evolution in *Lolium*. *Heredity, Lond.* **22**, 1–18.
Rees, H. & Narayan, R. K. J. 1977 Nuclear DNA variation in *Lathyrus*. In *Chromosomes today*, vol. 6 (ed. A. de la Chapelle & M. Sorsa), pp. 131–139. Amsterdam: Elsevier/North Holland Biomedical Press.
Rhoades, M. 1952 Preferential segregation in maize. In *Heterosis* (ed. J. W. Gowen), pp. 66–80. Ames: Iowa State College Press.
Rothfels, K., Sexsmith, E., Heimberger, M. & Krause, M. O. 1966 Chromosome size and DNA content of species of *Anemone* L. and related genera. *Chromosoma* **20**, 54–74.
Van't Hof, J. 1965 Relationships between mitotic cycle duration, S period duration and the average rate of DNA synthesis in the root meristem cells of several plants. *Expl Cell Res.* **39**, 48–58.

Discussion

P. R. DAY (*Plant Breeding Institute, Cambridge, U.K.*). In figure 1, Professor Rees indicated the range of DNA amounts found in cells of diploid species. What factors restrict the upper limit to about 200 pg?

H. REES. I can only make a guess. Cell volume is directly correlated with nuclear DNA amount. In some tissues it may be that there is an upper limit to the cell volume beyond which the diffusion of metabolites becomes inefficient. If so, there would be a constraint upon further increase in nuclear DNA.

… B 292, 579–588 (1981)

The analysis of plant genes and chromosomes by using DNA cloned in bacteria

By R. B. Flavell

Plant Breeding Institute, Maris Lane, Trumpington, Cambridge CB2 2LQ, U.K.

Some examples are described in which the molecular cloning of plant DNA and the associated techniques of molecular biology have already contributed to plant genetics and plant breeding research. They include (a) the isolation and determination of the complete sequence of some nuclear genes, (b) the detection of genetic variation in gene structure and copy number, (c) the physical mapping of the chloroplast genome, (d) the physical mapping and identification of nuclear chromosomes by *in situ* hybridization, (e) a molecular description of highly repeated sequences in heterochromatin, (f) a rapid method for classifying cytoplasmic variants of maize and (g) characterization of mitochondrial DNA sequences necessary for normal pollen development in maize. It is emphasized that the mapping of genomes by using molecular methods does not require genetic variation or expression of the genes in the phenotype. Where a nucleic acid probe is available for a gene, then recessive allelic variants or inactive genes can be detected that do not contribute to the phenotype.

Introduction

The introduction of the molecular cloning of DNA in bacteria and the associated techniques of molecular biology into plant genetics has greatly enhanced the exploration of genetic variation at the molecular level. Consequently, the application of the techniques of molecular biology to plant chromosomes may be seen as a milestone in the history of plant genetics and in the exploitation of variation in plant breeding. It is my aim here to describe briefly some examples where the application of molecular biology has already contributed new knowledge on the structure of plant genes and on the molecular nature of genetic variation. I also wish to emphasize that chromosomes and very small segments of chromosomes can be mapped by using molecular biochemistry without a requirement for phenotypically expressed variation or estimates of recombination frequencies. The approach is therefore different from classical genetic analysis. Molecular biological approaches sometimes lead to new, more rapid methods of detecting genetic variation. To illustrate this, examples of the analysis of DNA sequences relevant to the control of endosperm development in triticale and to pollen development in maize are described.

Molecular cloning has brought new opportunities for studies in plant genetics for two principal reasons. The first is that the insertion of a small fragment of plant DNA into a bacterial plasmid *in vitro*, and the transfer of this recombinant plasmid into a bacterium, isolates the plant DNA fragment from all the others of the plant cell because only one chimaeric plasmid is taken up by each bacterium. The second reason is that the subsequent growth of the recipient bacterium to form a clonal population of cells, from which the plant DNA can be reisolated, provides large quantities of the pure plant DNA sequence. With large quantities of a pure sequence, extensive studies, including nucleotide sequencing, can be carried out.

The limiting step in the isolation of genes by bacterial cloning is the detection of the bacterium carrying the desired plant gene. Detection is relatively straightforward when a nucleic acid sequence homologous to the gene is available, and this is why most of the genes isolated to date are those whose RNA products are in large enough concentration, in at least some plant tissues, to be easily purified or highly enriched. Such genes include those specifying the ribosomal RNAs, storage proteins and leghaemoglobin. Studies on the ribosomal RNA genes of wheat purified by molecular cloning, which have led to new knowledge on gene structure and variation, are described below.

Variation in gene structure and copy number uncovered by molecular analysis

The genes specifying the ribosomal 25S, 18S and 5S RNA molecules are highly repeated in higher plant genomes. Hybridization of purified 5S RNA to wheat DNA digested with a restriction endonuclease and fractionated by electrophoresis showed that the 5S RNA genes reside in two different classes of repeating unit, one about 410 base pairs long and the other about 500 base pairs long (Appels *et al.* 1980). Gerlach & Dyer (1980) have purified examples of each size of repeating unit by molecular cloning and determined their complete nucleotide sequence. Because the 5S RNA sequence had been determined previously (Payne & Dyer 1976; Barber & Nichols 1978), it was possible to recognize the 120 base pairs in each repeating unit that specify the 5S RNA. The remaining DNA in each repeating unit is 'spacer' DNA.

A particular feature of the sequenced member of the 500 base pair size class of repeating unit is that it contained a 15 base pair duplication within the coding sequence. It is not known how many copies of this mutant gene are in the wheat variety Chinese Spring, but approximately 15–20 % of the repeating units in the larger size class appeared to contain an additional 10–20 base pairs (Gerlach & Dyer 1980). Thus the 5S genes with the duplication may be common in this wheat variety. This is of particular significance because in *Xenopus* 5S RNA genes, the region homologous to that carrying the duplication in some wheat genes is very important in regulating transcription of the gene (Sakonju *et al.* 1980; Bogenhagen *et al.* 1980). No RNA transcript 135 (120+15) bases long was detected in wheat by Gerlach & Dyer (1980), so it appears that the duplication with the gene prevents transcription – an explanation consistent with the region carrying the duplication being important for gene expression also in wheat – or the RNA product is destroyed. Another possibility is that the genes with the internal duplication were silent in the tissues studied, owing to developmental control (not all 5S RNA genes are used at any one time).

In each repeat unit there is an array of A–T base pairs immediately following the sequence specifying the 5S RNA. This array dictates, in part, the termination of transcription. The 70 base pairs before the beginning of the coding sequence are very similar in the cloned sequences from both the 410 and 500 base pair size classes. Most of the other spacer DNA has diverged considerably between repeating unit classes. Thus the 70 base pairs before the start of the coding sequence are probably conserved to determine the correct initiation of transcription of 5S RNA.

This study on 5S RNA genes illustrates the value of molecular analysis to genetics in many ways. It has revealed (1) details of gene structure relevant to the control of gene expression, (2) variation in spacer DNA length and sequence between genes, (3) a small duplication within a gene, which may make the gene permanently silent thus preventing its presence ever being detected in studies of the phenotype. Also comparison of the nucleotide sequences of the 120 base pair regions coding for the 5S RNA in four clones has revealed single base pair differences (mutations) at eight different sites in different copies of the gene (Gerlach & Dyer 1980). The biological significance of these mutations is as yet unknown.

The copy number of repeated genes, such as the ribosomal RNA genes, commonly varies between individuals of a species. This is easily detected by using cloned sequences as probes. Variation in the copy number of 25S+18S ribosomal RNA genes which are tandemly arrayed at homologous loci in wheat and rye has been detected by *in situ* hybridization of cloned rDNA to metaphase chromosomes (Miller *et al.* 1980). Heterozygosity for the number of copies of rRNA genes has also been detected in diploid rye by similar methods. These rapid ways of detecting variation, including heterozygosity, without carrying out a breeding experiment, clearly illustrate the value of the techniques of molecular biology to genetics.

The physical mapping of chromosomes

The mapping of plant chromosomes by molecular techniques not based upon detecting variation in gene expression is best developed for the chloroplast genome (Bedbrook & Kolodner 1979). The progress obtained by this route is important not only because of the importance of the chloroplast genes to photosynthesis and other key areas of plant metabolism, but also because in many crop plants, recombination either does not occur between maternal and paternal chloroplast genomes or, if it does take place, it happens at a frequency below that necessary to enable the genes to be mapped by conventional genetic techniques. Segments of chloroplast DNA, generated by digesting purified DNA with restriction endonucleases, have been ordered on circular maps in a number of species including maize, pea, spinach (reviewed in Bedbrook & Kolodner 1979) and wheat (Bowman *et al.* 1981). Every segment has been cloned in bacteria, and consequently large quantities are available to map and sequence any region of the circular chromosome. The genes for chloroplast ribosomal RNAs and many transfer RNAs have been positioned on the genomes by hybridization of the RNAs to specific DNA fragments. The position of other genes, including that of the large subunit of ribulose bisphosphate carboxylase (LS) has been determined by finding the cloned DNA segment that, when mixed with appropriate extracts from *E. coli*, produces an mRNA that translates into LS (reviewed in Bedbrook & Kolodner 1979). The faithful transcription and translation of chloroplast genes in *E. coli* extracts implies that *E. coli* contains the information to recognize the start and stop signals for gene expression on chloroplast DNA. This has been recently proved by the demonstration that LS is synthesized *in vivo* by *E. coli* cells carrying the chloroplast LS gene (Gatenby *et al.* 1981). As expected from this result, the segment of DNA that occurs just before the coding region of the LS gene that has been cloned from maize possesses features similar to those that bacteria use to control the initiation of transcription and translation (McIntosh *et al.* 1980; Gatenby *et al.* 1981). From these examples it may be seen that the mapping of genes on chloroplast DNA is proceeding relatively rapidly, and detailed sequencing and other studies are providing details of the signals involved in gene expression. Such studies could scarcely be contemplated without resorting to molecular cloning.

Small segments of nuclear chromosomes, which, of course, contain much more DNA than the chloroplast genome, are being mapped by similar techniques. Furthermore, plant nuclear chromosomes contain many families of repeated sequences (Flavell 1980), and when members of the same family are highly clustered or in tandem, their position can be determined by *in situ* hybridization (Pardue & Gall 1970). In this technique, a particular piece of DNA purified by molecular cloning is tritium-labelled and hybridized to denatured metaphase chromosomes spread on a glass slide. The sites of hybridization are recognized by radioautography. The locations of the multicopy $25S + 18S$ cytosolic ribosomal RNA genes in wheat and rye have been revealed this way (Gerlach & Bedbrook 1979; Miller *et al.* 1980). This technique provides a way not only of recognizing the chromosomal sites of such purified genes, or other repeated sequences, but also of distinguishing the chromosomes containing them from those that do not. The physical mapping of chromosomes by *in situ* hybridization with the use of highly purified repeated sequence DNA is best developed in wheat and rye (Gerlach *et al.* 1978, 1980; Gerlach & Peacock 1980; Dennis *et al.* 1980; Jones & Flavell 1981 a, b; Bedbrook *et al.* 1980; Hutchinson *et al.* 1980) and further details for rye are described below.

A MOLECULAR ANALYSIS OF HIGHLY REPEATED SEQUENCES AND HETEROCHROMATIN IN RYE

The project to purify and study the chromosomal location of specific families of repeated sequences in rye was undertaken because (a) the heterochromatic segments at the telomeres of cultivated rye appear to be responsible in triticale for aberrations in meiosis (Merker 1976), endosperm development and grain shrivelling (reviewed in Thomas *et al.* 1980), and (b) heterochromatin contains mostly repeated sequence DNA (reviewed by John & Miklos 1979). The role of rye heterochromatin in causing aberrant endosperm development with consequent grain-shrivelling in triticale is described in the contribution to this symposium by Bennett.

The purification by molecular cloning of members of families of highly repeated sequences likely to reside in rye telomeric heterochromatin (Appels *et al.* 1978) is described in full elsewhere (Bedbrook *et al.* 1980). The organization of the families in the genome was discovered by carrying out *in situ* hybridization of the radioactively labelled cloned DNAs to metaphase chromosomes. Four families were found to account for 45–65 % of the DNA of heterochromatic C-bands (Bedbrook *et al.* 1980; Jones & Flavell 1981 a, b). In figure 1 the mapping of the four families of repeats in chromosomes of the rye variety King II is summarized. The C-bands, which are shown on the right hand member of each chromosome pair, are taken from the results of Singh & Röbbelin (1975, 1976) for King II.

The telomeric heterochromatin on the short arm of each chromosome includes tandem arrays of each of the four families. Chromosomes 1R, 4/7R, 5R and 6R contain interstitial heterochromatin bands that contain only one of the families, while chromosome 6R also contains interstitial arrays of another family. Where telomeric blocks of heterochromatin occur on the long arms, they include arrays of sequences from one, two or four of the families studied. Thus different blocks of heterochromatin can be distinguished by hybridization with different cloned sequences, enabling certain chromosomes to be distinguished one from another. However, each family of repeats is present on all chromosomes. Further structural details on the repeated sequences in these families are described elsewhere (Bedbrook *et al.* 1980). These details, together with the results summarized here (see also Jones & Flavell 1981 a, b), provide a molecular description of those segments of the genome that appear to be major contributors to grain-shrivelling in triticale, a phenomenon first observed by plant breeders.

It is interesting to note that a genetic analysis of grain-shrivelling in triticale would show that this phenomenon is controlled by many loci, some of larger effect than others. The inheritance would therefore be of the polygenic, continuous kind. However, the analysis of heterochromatin by molecular analysis shows that in this particular instance the polygenes are long arrays of repeated sequences that are probably not transcribed. It is possible that the arrays exert their phenotypic effect in triticale by being replicated later than most of the wheat and

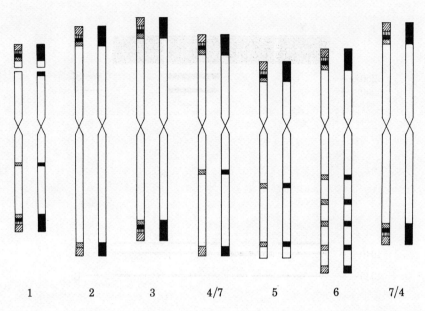

FIGURE 1. The composition of heterochromatic C-bands in rye, variety King II. The right-hand member of each chromosome pair illustrates the position of C-bands (Singh & Röbbelin 1975, 1976) and unpublished results of M. D. Bennett and colleagues). The left-hand member illustrates the arrays of repeated sequences found in the heterochromatin by *in situ* hybridization (Bedbrook *et al.* 1980; Jones & Flavell 1981 a). Molecular descriptions of each of the four kinds of array are given in Bedbrook *et al.* (1980). Each kind of repeating sequence has a different shading: ▨, ▥, ■, and ▩.

remaining rye DNA, with the consequence that rye anaphase bridges are formed but not resolved (Bennett 1977; Thomas *et al.* 1980). That is, the presence of the repeats affects the phenotype through their influence on chromosome behaviour in a specific tissue. This example shows that heterochromatin and highly repeated sequences are not inert when considered in the context of the total biology of an organism.

The cloned repeated sequences described here are, of course, useful probes for uncovering chromosomal variation within and between individuals. Many chromosomes have been found in *Secale* species with *in situ* hybridization patterns different from those illustrated in figure 1 (Jones & Flavell 1981 a, b). The probes are also useful together with C-banding for characterizing triticale lines in breeding programmes that have lost telomeric heterochromatin and are therefore likely to have less shrivelled grains (Bennett, this symposium).

The loss of telomeric highly repeated DNA from rye chromosomes maintained as disomic additions in wheat has already been demonstrated by the cloned probe DNAs (Jones & Flavell 1981 a).

584   R. B. FLAVELL

Variation in mitochondrial DNA and cytoplasmically inherited male sterility in maize

Techniques of molecular biology including molecular cloning have recently made useful contributions to our understanding and exploitation of cytoplasmic mutations that cause the failure of pollen development. These mutations are used extensively in hybrid crop breeding and hybrid seed production. For example, in maize during the 1960s almost all the hybrid

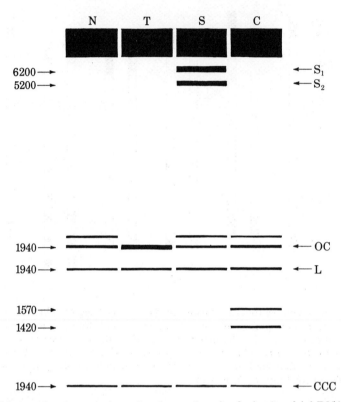

Figure 2. The classification of maize cytoplasms by electrophoresis of mitochondrial DNA. Mitochondrial DNA from each of the four types of cytoplasm (N, T, S and C) is fractionated by electrophoresis on an agarose gel. The linear and circular low molecular mass DNAs provide a unique pattern for each cytoplasm (results schematically redrawn from Kemble et al. 1980) and Kemble & Bedbrook 1980)). Electrophoresis was from the top to the bottom of the diagram. The major block of DNA at the top of the gel is high molecular mass DNA in the principal mitochondrial genome. S_1 and S_2 are linear pieces of DNA found only in S cytoplasm. OC, L and CCC are the open circular, linear and covalently closed circular forms respectively of the same nucleotide sequence. The numbers on the left-hand side are the lengths of the DNA pieces in nucleotide pairs.

maize seed production in the U.S.A. was achieved by the use of the 'Texas' cytoplasm, which causes severe pollen failure in certain nuclear genetic backgrounds under agricultural conditions (Duvick 1965). The pollen failure ensures that plants with the Texas cytoplasm are fertilized by pollen from another plant and therefore that only hybrid seed is formed. Male sterility due to cytoplasmically inherited mutations can be suppressed by specific nuclear genes (fertility restorer genes). This has enabled geneticists to distinguish three different types of maize cytoplasm (T, C and S) that confer male sterility. Each is suppressed by different restorer genes (Beckett 1971). To determine to which group an unknown cytoplasm belongs, it was formerly necessary to transfer it by backcrossing into the nuclear background of the tester stocks with different fertility restorer genes and then to score fertility. Cytoplasm classification can now be carried out much more rapidly by studying DNA components of mitochondria or the proteins synthesized in isolated mitochondria (Kemble et al. 1980; Forde et al. 1980).

Mitochondria of maize contain small pieces of DNA, in addition to their principal genomes. Some are linear, some are circular (Kemble & Bedbrook 1980). These small pieces are readily seen when mitochondrial DNA is fractionated by electrophoresis on an agarose gel and 'stained' with ethidium bromide. The complement of small DNA molecules is unique for each of the N, C, T and S cytoplasms, as illustrated in figure 2. This was concluded from a survey of 31 lines carrying different cytoplasms previously classified by their interaction with specific restorer genes (Kemble et al. 1980). The classification of maize cytoplasms by this method requires etiolated shoots from only 20–30 seeds, takes only 1 day after harvesting the plant material, many samples can be processed simultaneously and the characteristic gel banding pattern is independent of the nuclear genetic background, i.e. the cytoplasm can be classified into the appropriate group whether there is a nuclear restorer gene present or not. The method is therefore of considerable use to breeders for analysing sources of cytoplasmically inherited male sterility or other cytoplasmically inherited traits.

Progess in understanding the molecular basis of the variation causing infertility in plants with T, C or S cytoplasms has come from cloning DNA sequences in the two small DNA pieces, S_1 and S_2, found only in mitochondria from S cytoplasm (figure 2). Restriction endonuclease maps of these fragments, first recognized by Pring et al. (1977), have been established by research groups in Florida, North Carolina and Cambridge, U.K. Cloned fragments were selected that specifically hybridized only to S_1 or S_2 (Thompson et al. 1980) and used as hybridization probes to show that S_1 and S_2 sequences are present in the principal genome (high molecular mass mitochondrial DNA) in the normal cytoplasm but not in the male sterile cytoplasms S, T or C (Thompson et al. 1980). Thus male sterility is correlated with the deletion of S_1 and S_2 sequences from the mitochondrial genome. That the loss of S_2 sequences from the principal mitochondrial genome is the *cause* of cytoplasmic male sterility is strongly supported by the finding that phenotypic revertants of S cytoplasm, i.e. plants with normal pollen development, have lost the linear S_1 and S_2 molecules and have S_2 sequences integrated into high molecular mass mitochondrial DNA (Levings et al. 1980).

These studies therefore illustrate that a phenotypic characteristic important in plant breeding is probably the result of loss of DNA sequences or of the excision and maintenance of the sequences in a small, presumably non-functional fragment. The reversion of the phenotype upon reintegration of the sequences suggests that the location of the sequences is important for correct function. Furthermore, the S_1 and S_2 cloned DNA fragments are very useful probes for detecting plants with the deletion(s). Such plants may possess a cytoplasm likely to produce pollen failure in certain environments and nuclear genetic backgrounds. This assay, which would give a positive answer whether nuclear restorer alleles were present or not, again illustrates the value of molecular biological approaches for detecting recessive variation.

Prospects

Examples such as those described in this paper represent the beginning of the application of molecular cloning to plant genetics. A great deal of new information about genetic variation, gene expression, differentiation and development can be confidently expected to emerge over the next few years. If genetic problems important in plant breeding are studied, such as that described above relating to cytoplasmic male sterility, there is every chance that information and techniques helpful to plant breeders will emerge. But even more than this, when it becomes possible to insert DNA into plants, the genes isolated by molecular cloning techniques can be used to modify plants in novel directed ways (Cocking 1981; Flavell 1981). This achievment will herald another era in the history of plant genetics and breeding made possible by molecular cloning. There is indeed much to look forward to in plant genetics because of the results from the past 40 years of intensive research in bacterial genetics.

I am grateful to my colleagues Dr T. A. Dyer and Dr C. N. Law for help in the preparation of the manuscript and to other colleagues at the Plant Breeding Institute who provided much of the experimental evidence presented in this paper.

References (Flavell)

Appels, R., Driscoll, C. & Peacock, W. J. 1978 Heterochromatin and highly repeated DNA sequences in rye (*Secale cereale*). *Chromosoma* **70**, 67–89.

Appels, R., Gerlach, W. L., Dennis, E. S., Swift, H. & Peacock, W. J. 1980 Molecular and chromosomal organisation of DNA sequences coding for the ribosomal RNAs in cereals. *Chromosoma* **78**, 293–311.

Barber, C. & Nichols, J. L. 1978 Conformation studies on wheat embryo 5S RNA using nuclease S_1 as a probe. *Can. J. Biochem.* **56**, 357–364.

Beckett, J. B. 1971 Classification of male-sterile cytoplasms in maize (*Zea mays* L.). *Crop. Sci.* **11**, 724–727.

Bedbrook, J. R. & Kolodner, R. 1979 The structure of chloroplast DNA. *A. Rev. Pl. Physiol.* **30**, 593–620.

Bedbrook, J. R., Jones, J., O'Dell, M., Thompson, R. D. & Flavell, R. B. 1980 A molecular description of telomeric heterochromatin in *Secale* species. *Cell* **19**, 545–560.

Bennett, M. D. 1977 Heterochromatin, aberrant endosperm nuclei and grain shrivelling in wheat-rye genotypes. *Heredity, Lond.* **39**, 411–419.

Bogenhagen, D. F., Sakonju, S. & Brown, D. D. 1980 A control region in the center of the 5S RNA gene directs specific initiation of transcription. II. The 3' border of the region. *Cell* **19**, 27–35.

Bowman, C. N., Koller, B., Delius, H. & Dyer, T. A. 1981 Physical mapping of wheat chloroplast DNA. *Molec. gen. Genet.* (Submitted.)

Cocking, E. C. 1981 Plant genome manipulations. *Nature, Lond.* (In the press.)

Dennis, E. S., Gerlach, W. L. & Peacock, W. J. 1980 Identical polypyrimidine–polypurine satellite DNAs in wheat and barley. *Heredity, Lond.* **44**, 349–366.

Duvick, D. N. 1965 Cytoplasmic pollen sterility in corn. *Adv. Genet.* **13**, 1–56.

Flavell, R. B. 1980 The molecular characterisation and organisation of plant chromosomal DNA sequences. *A. Rev. Pl. Physiol.* **31**, 569–596.

Flavell, R. B. 1981 Needs and potential of molecular genetic modification in plants. In *Proc. Rockefeller Workshop*, Rockefeller Foundation, New York. (In the press.)

Forde, B. G., Oliver, R. J. C. & Leaver, C. J. 1980 Classification of normal and male sterile cytoplasms in maize. I. Electrophoretic analysis of variation in mitochondrically synthesised proteins. *Genetics, Princeton* **95**, 443–450.

Gatenby, A. A., Castleton, J. A. & Saul, M. J. 1981 Expression in *E. coli* of the maize and wheat chloroplast genes for the large subunit of ribulose bisphosphate carboxylase. *Nature, Lond.* (In the press.)

Gerlach, W. L. & Bedbrook, J. R. 1979 Cloning and characterisation of ribosomal RNA genes from wheat and barley. *Nucl. Acid Res.* **7**, 1869–1886.

Gerlach, W. L. & Dyer, T. A. 1980 Sequence organisation of the repeating units in the nucleus of wheat which contain 5S rRNA genes. *Nucl. Acid Res.* **8**, 4851–4865.

Gerlach, W. L. & Peacock, W. J. 1980 Chromosomal locations of highly repeated DNA sequences in wheat. *Heredity, Lond.* **44**, 269–276.

Gerlach, W. L., Appels, R., Dennis, E. S. & Peacock, W. J. 1978 Evolution and analysis of wheat genomes using highly repeated DNA sequences. In *Proc. 5th Int. Wheat Genet. Symp.* (ed. S. Ramanujam) vol. 1, pp. 81–91.

Gerlach, W. L., Miller, T. E. & Flavell, R. B. 1980 The nucleolus organiser of diploid wheats revealed by *in situ* hybridisation. *Theor. appl. Genet.* **58**, 97–100.

Hutchinson, J. H., Miller, T. E. & Chapman, V. 1980 Chromosome pairing at meiosis in hybrids between *Aegilops* and *Secale* species: a study by *in situ* hybridisation using cloned DNA. *Heredity, Lond.* **45**, 245–254.

John, B. & Miklos, G. L. 1979 Functional aspects of satellite DNA and heterochromatin. *Int. Rev. Cytol.* **58**, 1–114.

Jones, J. D. G. & Flavell, R. B. 1981a The mapping of highly repeated DNA families and their relationship to C bands in chromosomes of *Secale cereale*. *Chromosoma*. (Submitted.)

Jones, J. D. G. & Flavell, R. B. 1981b Repeated DNA sequences in the genus *Secale*. *Chromosoma*. (Submitted.)

Kemble, R. J. & Bedbrook, J. R. 1980 Low molecular weight circular and linear DNA in mitochondria from normal and male sterile *Zea mays* cytoplasm. *Nature, Lond.* **284**, 565–566.

Kemble, R. J., Gunn, R. E. & Flavell, R. B. 1980 Classification of normal and male-sterile cytoplasms in maize. II. Electrophoretic analysis of DNA species in mitochondria. *Genetics, Princeton* **95**, 451–458.

Levings, C. S., Kim, B. D., Pring, D. R., Conde, M. F., Mans, R. J., Laughnan, J. R. & Gabay-Laughnan, S. J. 1980 Cytoplasmic reversion of cms-*S* in maize: association with a transpositional event. *Science, N.Y.* **209**, 1021–1023.

McIntosh, L., Poulsen, C. & Bogorad, L. 1980 The DNA sequence of a chloroplast gene: the large subunit of ribulose bisphosphate carboxylase of maize. *Nature, Lond.* **288**, 556–560.

Merker, A. 1976 The cytogenetic effect of heterochromatin in hexaploid triticale. *Hereditas* **83**, 215–222.

Miller, T. E., Gerlach, W. L. & Flavell, R. B. 1980 Nucleolus organiser variation in wheat and rye revealed by *in situ* hybridisation. *Heredity, Lond.* **45**, 377–382.

Pardue, M. L. & Gall, J. 1970 Chromosomal localisation of mouse satellite DNA. *Science, N.Y.* **168**, 1356–1358.

Payne, P. I. & Dyer, T. A. 1976 Evidence for nucleotide sequence of 5S rRNA from the flowering plant *Secale cereale* (rye). *Eur. J. Biochem.* **71**, 33–38.

Pring, D. R., Levings, C. S., Hu, W. W. L. & Timothy, D. H. 1977 Unique DNA associated with mitochondria in the 'S' type cytoplasm of male sterile maize. *Proc. natn. Acad. Sci., U.S.A.* **74**, 2904–2908.

Sakonju, S., Bogenhagen, D. F. & Brown, D. D. 1980 A control region in the centre of the 5S RNA gene directs specific initiation of transcription. I. The 5' border of the region. *Cell* **19**, 13–25.

Singh, R. J. & Röbbelin, G. 1975 Comparison of somatic giemsa banding pattern in several species of rye. *Z. PflZücht.* **75**, 270–280.

Singh, R. J. & Röbbelin, G. 1976 Giemsa banding technique reveals deletions within rye chromosomes in Addition lines. *Z. PflZücht.* **76**, 11–18.

Thomas, J. B., Kaltsikes, P. J., Gustafson, J. P. & Ronpakias, D. G. 1980 Development of kernel shrivelling in triticale. *Z. PflZücht.* **85**, 1–27.

Thompson, R. D., Kemble, R. & Flavell, R. B. 1980 Variations in mitochondrial DNA organisation between normal and male sterile cytoplasms of maize. *Nucl. Acid Res.* **8**, 1999–2008.

Discussion

P. R. Day (*Plant Breeding Institute, Cambridge, U.K.*). Most plant breeders will be daunted by the detailed analysis necessary to identify and manipulate useful genes in plant transformation. In the meantime, what are the prospects of introducing useful variation by the cruder techniques of transforming with shotgun plant DNA clones?

R. B. Flavell. I cannot at present be particularly optimistic about 'shotgun' approaches to plant modification because they are so inefficient. Consider a plant with a haploid genome size about that of barley (5.3×10^9 nucleotide pairs). If a bank of cloned sequences were created, with the average piece of plant DNA being 20 000 nucleotide pairs, then approximately 2.6×10^5 different clones would be required to provide a complete complement of the genome. Clearly in practice the number of clones required would be substantially in excess of this. If only one cloned DNA piece were taken up into each plant, any single copy gene would be present in only one out of 260 000 plants. Because in practice the efficiencies would be *very much* less than this for technical as well as theoretical reasons, the number of plants necessary to be handled is too great to make the approach attractive. However, if techniques of specific gene enrichment are developed and/or important genes or regulatory sequences are represented many times in a plant genome, the shotgun approach becomes more favourable. Clearly, unless transformation frequencies approach 100% it is also essential to be able to select those plants containing inserted DNA.

J. HESLOP-HARRISON, F.R.S. (*Welsh Plant Breeding Station, Aberystwyth, U.K.*). Although there is still some uncertainty about what cytoplasmic organelles, if any, are actually conveyed into the egg in the course of fertilization in angiosperms, some aspects are fairly clear. Notwithstanding some earlier suggestions to the contrary, it is now well established that the plastid and mitochondrial lineages are continuous through meiosis in the anther, although the organelles do undergo dedifferentiation followed by later redifferentiation during the meiotic prophase (H. G. Dickinson & J. Heslop-Harrison, *Phil. Trans. R. Soc. Lond.* B **277**, 327–342 (1977)). During the subsequent pollen mitosis, the organelles are partitioned between the vegetative and generative cells; mitochondria appear invariably to pass into the generative cell, but plastids may (e.g. in Gramineae) or may not (e.g. in Orchidaceae) be incorporated. Where plastids do not enter the generative cell, there seems no likelihood that any can reach the egg. There is still a further period when there could be an elimination of organelles, namely during the passage through the receptive synergid. At this time the male gametes appear to be thoroughly 'scrubbed' of pollen tube cytoplasm, so there is little likelihood of organelles from this source passing into the egg. Light microscopic images often appear to show naked male nuclei participating in the two fertilizations, and indeed some electron microscopic evidence (e.g. by D. D. Cass & W. A. Jensen, *Am. J. Bot.* **57**, 62–70 (1970)) might be taken to indicate that the gametes are themselves cleansed of organelles in passing through the synergid. However, this does not appear to be convincingly established yet for any species. The genetical evidence, of course, indicates that in at least some families, plastids must be transferred into the egg with the male gamete. I would expect there to be variation in this, just as there is in the earlier partition of organelles after pollen mitosis.

The molecular and genetic manipulation of nitrogen fixation

By J. R. Postgate, F.R.S., and F. C. Cannon

A.R.C. Unit of Nitrogen Fixation, University of Sussex, Falmer, Brighton, BN1 9RQ, U.K.

The fundamental importance of dinitrogen fixation for world agriculture, in relation to projected energy supplies, population pressure and food requirements over the next decades, obliges scientists to reconsider ways of exploiting this biological process. Genetic manipulation offers several options in principle. Existing symbiotic systems such as the legumes and seemingly inefficient systems such as the grass associations could be improved; new symbioses could be developed by *nif* gene transfer to rhizosphere commensals or by somatic hybridization of appropriate plants. A major advance would be to render plants independent of microbes by manipulation of expressable *nif* into the plant genome. This goal is discussed. It requires the complete genetic and physical characterization of *nif*, in particular its regulation, and an understanding of the physiological background within which *nif* can be expressed, as well as the ability to fuse *nif* to alien genetic systems. Substantial progress in these directions has been made by using the *nif* genes of *Klebsiella pneumoniae*; this progress is reviewed. Strategies for the further manipulation of *nif* towards regulated expression in the plant genome are considered.

Nitrogen fixation

The transcendent importance of nitrogen fixation in sustaining the biosphere has been recognized for much of the present century. In the last two or three decades the world's human population has outstripped the ability of natural nitrogen fixation processes, spontaneous and biological, to support adequate food production, so that more than 30 % of the world's population now depends on artificial N fertilizer for its minimal nutrition. Documentation of the quantitative and scientific bases of these assertions is widely available (see, for example, Postgate (1980*a*, *b*) and references therein). Essentially, nitrogen fixation returns to the biosphere nitrogen atoms that have been lost into the atmosphere as a result of denitrification. Both nitrogen fixation and denitrification are primarily biological processes, though chemical processes supplement them on a global scale, and both are exclusive to prokaryotic microbes: the bacteria. (One instance of nitrogen fixation by a eukaryotic microbe has been reported by Yamada & Sakaguchi (1980), but the physiology and taxonomic status of this interesting green alga is not clear.)

Energy and transport costs of the Haber–Bosch process, as well as threats of environmental catastrophe from greatly augmented use of chemical N fertilizer make further exploitation of chemical N fertilizer an unattractive solution to world food problems. This conclusion has excited much interest in the biological process, and many proposals for more extensive exploitation of biological nitrogen fixation have been presented. Though N fertilizer use is inescapable, and greater exploitation of existing diazotrophic systems will certainly come about, even together they will not solve the problem of feeding the human population of the world by the early decades of the next millennium; more imaginative developments, involving genetic and somatic manipulation of existing and newly developed systems, will be necessary (see Postgate (1980*b*) for references to this discussion).

Options

For this contribution we shall restrict our presentation to the improvement of existing systems involving plants, or the development of new plant systems, and shall say nothing of the possible exploitation of microbes *per se*, of the nitrogenase enzyme or of chemical catalysts based thereon.

Improved systems

Legumes, which form diazotrophic associations with bacteria of the genus *Rhizobium*, are the most important of the agricultural crops that are capable of exploiting biological nitrogen fixation. Plants such as peas, pulses, beans, lucerne and clover can make substantial contributions to food and fibre production with minimum N fertilizer costs as long as the plant and environment are appropriate. Other diazotrophic associations are exploited to a lesser extent. The efficacy of the *Rhizobium*–legume association, and in principle of all diazotrophic associations, depends on a variety of factors, some of which are amenable to genetic manipulation. A few illustrative examples follow.

Competitiveness

'Wild' rhizobia are often relatively ineffective symbionts in the sense that they transfer little fixed N to the plant, yet they often out-compete more effective strains which have been introduced exogenously. Genetic manipulation (e.g. transfer of effectiveness genes to highly competitive strains) would improve this situation.

Hydrogenase

Possession of an uptake hydrogenase by the rhizobia augments the efficiency of the soybean symbiosis by a mechanism that is partly understood (see Evans *et al.* 1980). Mutants in the genes for hydrogenase (Hup^- mutants) are available and the manipulation of these genes into strains of rhizobium with other virtues will become possible. Some species of *Rhizobium* (e.g. *R. leguminosarum*; Ruiz-Argueso *et al.* 1978) seem always to be deficient in hydrogenase, a situation that ought to be genetically correctable.

Fixation period

In many legumes, nitrogen fixation ceases before pod-filling, and manipulation of the plant or bacteria to prolong the fixation period would enhance crop yields (Hardy 1976).

The plant host

Breeding of plants for selectivity towards effective rhizobia and/or for enhanced uptake of CO_2 (and transfer of photosynthate to the symbiont) are examples of prospects for genetic manipulation of the plant partner of the legume symbiosis (Hardy 1976).

Cereal associations

Diazotrophic associations of *Azotobacter* (Dobereiner *et al.* 1972), *Azospirillum* (Dobereiner & Day 1976), *Bacillus* (Larson & Neal 1978) and other diazotrophs (Barber *et al.* 1978) with the roots of grasses or cereals are well established. Such diazotrophic rhizocoenoses (Dobereiner &

De-Polli 1980) are relatively ineffective, but genetic manipulation of both plant and microbe ought to yield agronomically more important associations.

Extensive discussion of the possibilities for generating improved systems, which should cover actinomycete and cyanobacterial associations (see Sprent 1979) as well as those mentioned above, is inappropriate to this paper.

New systems

The generation of new diazotrophic systems has been an attractive long-term proposition since the demonstration of intergeneric *nif* gene transfer among enteric bacteria (Dixon & Postgate 1972), and several new species of diazotrophic bacteria have been generated by conjugational transfer of laboratory-constructed *nif* plasmids (see Krishnapillai & Postgate 1980). Generation of new symbioses is a far more complex process and may be approached in three major ways.

1. Transfer of *nif*, and any other ancillary genes required for regulated expression, into organisms that already have close mutualistic associations with plants. In fact, such associations often exist where the selection pressure is favourable, as in the grass rhizocoenoses mentioned above. Transfer of expressible *nif* to mycorrhiza is an interesting possibility of this kind (see Giles & Whitehead 1975), but requires sustained and regulated expression of essentially prokaryotic genes in a eukaryotic background; this problem ought not to be insurmountable.

2. Somatic hybridization (see Cocking, this symposium) of appropriate plants to generate new hybrids that combine agricultural desirability with diazotrophic capability. Such projects do not involve genetic manipulation of *nif* and we shall not discuss them further.

3. Introduction of expressible *nif* into the plant genome. This approach, if successful, would have the important advantage that the plant would not only become diazotrophic in its own right, like a highly effective diazotrophic symbiosis, but it would also become independent of a prokaryotic symbiont, and thus unconcerned by problems of effectiveness and competitiveness which complicate the husbandry of existing diazotrophic associations.

This third approach can be subdivided. One plan would be to seek a benign prokaryotic endosymbiont capable of diazotrophy which would establish itself as a new diazotrophic organelle-like structure, analogous to a chloroplast. We are unable to prescribe a systematic strategy for the construction of a stable, regulated diazotrophic organelle so we shall not discuss this matter further. The transfer of *nif* genes from a diazotroph, together with ancillary genes, into some pre-existing part of the plant genome can be approached systematically. It is a long-term project, since it absolutely requires a solution to the problem of expression of a complex of prokaryotic genes in a eukaryote, but it is also the most challenging, since present knowledge of biological nitrogen fixation indicates no reason why it should not be successful. Before discussing strategies for achieving this option, we shall briefly indicate the present state of knowledge of the *nif* genes, their expression and manipulation.

Progress

Nif *genes in* Klebsiella pneumoniae

The close linkage of *his* and *nif* on the *K. pneumoniae* chromosome facilitated the construction of self-transmissible plasmids that carry the *his–nif* region (Cannon *et al.* 1976; Dixon *et al.* 1976). The plasmid pRD1 has the genetic properties of its precursor, the P-incompatibility group plasmid RP4, and also carries the chromosomal genes *gnd rfb his nif shi*A. This plasmid and several derivatives have been used for complementation analyses and fine structure mapping of *nif* mutations. pRD1 has also been used to investigate the expression of *Klebsiella nif* genes in other bacterial genera (Dixon *et al.* 1976; Cannon & Postgate 1976; Postgate & Krishnapillai 1977; Krishnapillai & Postgate 1980) and was the source of DNA for cloning the *nif* genes.

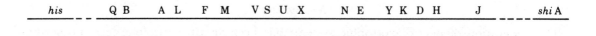

FIGURE 1. Map of *Klebsiella pneumoniae nif* genes (MacNeil *et al.* 1978; Merrick *et al.* 1980; Riedel *et al.* 1979; Puhler & Klipp 1981). The arrows indicate *nif* operons and their directions of transcription.

Seventeen contiguous *nif* genes have been identified and mapped close to the *his* operon (figure 1). Fourteen of these were recognized by complementation analysis, and further genetic characterization of mutations from one of the complementation groups suggested that it was composed of two genes, *nif*A and L (MacNeil *et al.* 1978; Merrick *et al.* 1980). This was confirmed by the identification of the polypeptides specified by *nif*A and *nif*L in strains carrying *nif*AL gene clones (Puhler & Klipp 1981; F. C. Cannon, unpublished). The gene order, *his*... *nif*QBALFMVSUNEKDHJ was unambiguously established by deletion mapping (MacNeil *et al.* 1978; Merrick *et al.* 1980). Two additional genes, *nif*Y, located between *nif*E and K, and *nif*X, located between *nif*U and N (see figure 1), were identified on the basis of polypeptide elimination in *Escherichia coli* minicells carrying *nif* gene clones with Tn5 insertions. The locations of *nif*X and Y were determined by physical mapping of the Tn5 insertions (Puhler & Klipp 1981). Complementation analysis of insertion mutations that exert transcriptional polarity indicated that the *nif* genes were arranged in seven or eight operons and that the multicistronic operons were transcribed in the same direction as the *his* operon (figure 1). RNA–DNA hybridization studies also showed that *his*-specific and *nif*-specific mRNA synthesized *in vivo* at detectable levels under the conditions used were transcribed from the same DNA strand (Janssen *et al.* 1980).

Nif *gene clones*

The location of *nif* genes in a single cluster linked to *his* on the plasmid pRD1 facilitated their molecular cloning by the construction of a series of small plasmids carrying overlapping DNA restriction fragments that collectively cover the entire *nif* region (Cannon *et al.* 1977, 1979). Puhler *et al.* (1979*a b*), using a different strategy, have also constructed clones that carry all the genes of the *nif* cluster, and MacNeil & Brill (1980) have isolated *in vivo* a series of λ*nif* transducing phages, which collectively carry the complete cluster of *nif* genes.

The sites at which several restriction endonucleases cleave the *nif* region were mapped by using the *nif* gene clones and this allowed the assignment of *nif* insertion mutations (induced by translocatable genetic elements) to physical locations with respect to the restriction sites (Cannon *et al.* 1979; Riedel *et al.* 1979). The physical *nif* map derived from these studies is in complete agreement with the genetic map. A more detailed physical map of the *nif* genes and an unambiguous identification of most of their products has been obtained by mapping the sites of Tn5 insertions derived by 'saturation mutagenesis' of *nif* gene clones and monitoring any resulting changes in number and/or size of *nif*-specific polypeptides synthesized in *E. coli* minicells (Puhler & Klipp 1981). The size of the *nif* region is approximately 23×10^3 base pairs and the results of physical mapping suggest that there are no gaps in the current *nif* map.

TABLE 1. Nif GENE PRODUCTS

nif gene	$10^{-3} M_r$ of gene product	gene function
Q	?	?
B	?	FeMoco synthesis or processing
A	57†	regulation
L	45†	regulation
F	17‡	electron transport
M	28§	processing Kp2
V	42§	processing Kp1
S	45‡	?
U	32, 25§	?
X	18§	?
N	50‡	FeMoco synthesis or processing
E	46‡	FeMoco synthesis or processing
Y	21§	regulation?
K	60‡	β-subunit of Kp1
D	56‡	α-subunit of Kp1
H	35‡	subunit of Kp2
J	120‡	electron transport

References for identification of gene products are: †, F. C. Cannon (unpublished); ‡, Roberts et al. (1978); §, Puhler & Klipp (1981).

Nif *gene products*

The molecular masses and probable functions of most *nif* gene products are listed in table 1. *K. pneumoniae* nitrogenase is composed of two redox proteins (for review see Mortenson & Thorneley 1979). The Fe protein, also termed Kp2 (M_r 68000) is a dimer composed of two identical subunits specified by *nif*H (Roberts et al. 1978). It contains a single Fe_4S_4 cluster and acts as an electron donor to the MoFe protein (Kp1) (M_r 218000). Kp1 contains 2 Mo atoms and approximately 32 Fe atoms, some of which are present as Fe_4S_4 clusters. Kp1 is a tetramer of two non-identical subunits (α and β, M_r 56000, 60000 respectively), which are the products of *nif*D (α) and *nif*K (β). The products of *nif*HDK alone cannot generate active nitrogenase and there is now considerable evidence that suggests post-translational processing or modification of the nascent polypeptides by other *nif* gene products.

A low molecular mass cofactor (FeMoco), which can be extracted from Kp1 and may contain the active site of the enzyme, partly restores activity to nitrogenase in cell extracts of *nif*B, *nif*E and *nif*N mutants (Roberts et al. 1978). These mutants are probably defective in the synthesis of FeMoco and/or its insertion into Kp1. Polypeptides with relative molecular masses of 46000 and 50000 have been assigned to *nif*E and *nif*N respectively (Roberts et al. 1978). Nitrogenase activity, which is impaired in *nif*M mutants, is partly restored in cell extracts of these mutants by the addition of purified Kp2, indicating that the product of *nif*M may play a role in the maturation of Kp2 (Roberts et al. 1978). The product of *nif*V is probably involved in processing Kp1, since purified Kp1 from a *nif*V mutant had altered substrate reduction properties (P. McLean & R. A. Dixon, unpublished).

The products of *nif*F (M_r 17000) and probably *nif*J (M_r 120000) are involved in the electron transport system for nitrogenase. The low levels of nitrogenase activities *in vivo* in *nif*F and *nif*J mutants are significantly enhanced in cell extracts assayed *in vitro* with sodium dithionite as an electron donor for nitrogenase (Roberts et al. 1978). Pyruvate can also act as an electron donor to nitrogenase in wild-type extracts but not in extracts of *nif*F and *nif*J mutants. Pyruvate-supported activity can be restored in extracts of *nif*F (but not *nif*J) mutants by the addition of *Azotobacter chroococcum* flavodoxin, which suggests that the product of *nif*F is an electron-transport protein (Hill & Kavanagh 1980; Nieva-Gomez *et al.* 1980).

The phenotypes of *nif*A and *nif*L mutants suggest that their products have regulatory functions. The product of *nif*A is required for the expression of all *nif* genes (except *nif*A and *nif*L) (Dixon *et al.* 1980; MacNeil *et al.* 1978; Roberts *et al.* 1978). The role of the *nif*A product as a positive activator of *nif* transcription is supported by the results of complementation tests in which the product of *nif*A can activate *nif* derepression in *trans* (Dixon *et al.* 1977; MacNeil *et al.* 1978). A specific regulatory function has not yet been assigned to the product of *nif*L, although there is evidence to suggest that it is involved in repression of the *nif* gene cluster (S. Hill, C. Kennedy & M. Merrick, unpublished). The *nif*Y gene product may also have a regulatory role since it appears to be required for full expression of the *nif*HDKY operon (M. C. Cannon, unpublished). The functions of the remaining genes, *nif*Q, S, U, X are unknown at present.

Regulation of K. pneumoniae nif *genes*

Ammonia and other forms of fixed nitrogen, including nitrate and amino acids, prevent the expression of nitrogenase activity. Vigorous aeration also blocks nitrogenase expression (St John *et al.* 1974; Eady *et al.* 1978). N_2 is not required as an inducer of *nif* expression, since *nif* is expressed in cultures grown under argon (Parejko & Wilson 1970). Therefore the expression of *nif* is mediated by depression rather than an induction mechanism.

Regulatory studies suggest that *nif* derepression in limiting ammonia concentrations is initiated by a protein or proteins involved in glutamine synthetase (Gs) expression, which appears to act as a transcriptional activator of *nif* (Streicher *et al.* 1974; Ausubel 1979). The same regulatory protein(s) are required for the expression of several other operons, including *hut* (histidine utilization), involved in the assimilation of fixed nitrogen sources (Tyler 1978). At least one of these regulatory genes is closely linked to *gln*A, the structural gene for GS (Streicher *et al.* 1974; Ausubel *et al.* 1979; Pahel & Tyler 1979; Kustu *et al.* 1979; G. Espin & M. Merrick, unpublished).

A model for *nif* derepression that is consistent with the available evidence is that the *gln*-gene controlling protein(s) act at a single site in the *nif* cluster, i.e. the promoter of the *nif*LA operon, to activate *nif*A transcription. The product of *nif*A in turn activates transcription of the other *nif* operons. Results consistent with this model are briefly summarized as follows: (1) *nif*A mutations are pleiotropic on the other *nif* operons (Dixon *et al.* 1977; MacNeil *et al.* 1978; Roberts *et al.* 1978); (2) *gln*A-linked mutations can give rise to either a partial *nif* constitutive or a Nif⁻ phenotype (Streicher *et al.* 1974; Tubb 1974; Ausubel *et al.* 1979; G. Espin & M. Merrick unpublished); (3) *cis*-dominant *nif* mutations have been isolated that are independent of *gln*-mediated regulation (Ausubel *et al.* 1977) (these mutations, called *nif*T, map in the *nif*L–A region (C. Kennedy, unpublished)); (4) although the product of *nif*A is required for the expression of the other *nif* operons it is not required for its own synthesis (Dixon *et al.* 1980; F. C. Cannon, unpublished); (5) gene fusions in which the *E. coli lac* operon was fused to individual *nif* promoters have been used to investigate derepression of *nif* operons independently of nitrogenase activity (Dixon *et al.* 1980). The results of these studies show that derepression of all *nif* operons is coordinate.

All identifiable *nif* gene products are absent in ammonia-grown cultures (Eady *et al.* 1978; Roberts *et al.* 1978) and *nif* mRNA is undetectable by DNA–RNA hybridization in repressed cultures (Janssen *et al.* 1980). *Nif–lac* fusions have also been used to show that all *nif* operons are repressed by ammonia, although the level of repression varies among different fusions (Dixon *et al.* 1980). When ammonia is added to a fully derepressed culture, a rapid repression of the Nif phenotype occurs (Tubb & Postgate 1973) and both nitrogenase activity and polypeptides disappear within 40 min (Collmer & Lamborg 1976). Studies on the repression of *nif*A and L gene clones suggest that the rapid repression of the *nif* operons by ammonia is not regulated by the removal of the *nif*A product (F. C. Cannon, unpublished). The results of these experiments show that although the rate of *nif*A and *nif*L polypeptide synthesis is proportional to the concentration of ammonia present in the culture during derepression, they are synthesized at ammonia concentrations that completely repress the nitrogenase structural gene operon. This suggests that rapid ammonia repression is mediated either through an attenuator mechanism or by an allosteric repressor. Repression studies with *nif*L mutants, which show that they are less sensitive to *nif* repression by O_2 (Dixon *et al.* 1980; S. Hill & C. Kennedy, unpublished) and fixed nitrogen (M. Merrick, unpublished), suggest that the *nif*L product has a repressor function. A model for ammonia repression of *nif* that is consistent with the evidence summarized above is that immediate repression of *nif* by the addition of ammonia to a derepressed culture is mediated through *nif*L product. The repression of GS by ammonia leads, in turn, to repression of the *nif*LA operon and thus concomitant removal of the *nif* repressor and activator. Repression of *nif* in ammonia-grown cultures would therefore be maintained solely by the absence of *nif*A product.

Studies with *nif*::*lac* fusions provide evidence that the nitrogenase structural gene operon in contrast to the other *nif* operons is autogenously regulated (Dixon *et al.* 1980). The presence of molybdate and product(s) of the *nif*HDKY operon is required for maximum expression of this operon. Although there is evidence to suggest that the *nif*Y gene product is required for maximum expression from the *nif*H promoter, it has not yet been assigned a specific regulatory function.

Strategies

The possibilities for genetic manipulation in animal cell cultures, in yeast strains and in several bacterial genera have increased dramatically with the recent development of transformation procedures and of a variety of gene vectors appropriate to each system. Studies of foreign gene expression in these systems are no longer hindered by problems of exogenous DNA uptake and maintenance. The type of vector used for the introduction of these genes usually determines whether they are maintained on autonomously replicating plasmids or by chromosomal integration. Vectors used for yeast transformation provide some illustrative examples. An indigenous 2 μm yeast plasmid and a centromere fragment of chromosome 6 have been used in the construction of two distinct groups of gene vectors that replicate autonomously in *Saccharomyces cerevisiae* (Fink 1980). A third group of vectors that promote chromosomal integration is derived from *E. coli* gene vectors carrying yeast markers such as *leu*, *his* and *ura*. The latter provide homology with the chromosome of origin and usually determine the site of integration for linked genes.

Although efficient procedures for the transformation of plant cells have not yet been developed some success has been achieved by using cauliflower mosaic virus (CaMV) DNA and Ti plasmid DNA from *Agrobacterium tumefaciens* (Hohn *et al.* 1980; Hernalsteens *et al.* 1980; Cocking, this symposium). Both of these DNA molecules are potentially useful for vector construction. Vectors similar to those based on the 2 μm yeast plasmid could be derived from the CaMV DNA. Since the T-DNA region of the Ti plasmid is known to integrate into plant chromosomes, it is possible that a vector carrying this region could promote integration of linked genes. The origins of replication from mitochondrial, chloroplast and plant chromosomal DNA are also potentially useful starting materials for vector construction. Progress in the construction of vectors is seriously hindered at present by the lack of useful plant selectable markers.

Although the availability of gene vectors is a prerequisite for the transfer of *nif* genes to plant cells, developing mechanisms for their expression will probably be a more challenging problem to resolve because of this organizational and regulatory complexity. An example is the recent transfer without expression of *nif* to *S. cerevisiae*, which was achieved with clones of the entire *nif* gene cluster (Elmerich *et al.* 1980). If these clones were transferred to *Schizosaccharomyces pombe* it is possible that at least transcription of the regulatory *nif*LA operon would be activated by protein(s) involved in the expression of glutamine synthetase in this yeast that has characteristics similar to those of the *Klebsiella gln* system (Van Ardel & Brown 1977).

The location into which *nif* genes are manipulated in plant cells will determine the type of regulatory sequences required for their expression. A chromosomal location would probably require the fusion of each essential *nif* gene to a eukaryotic promoter. A promotor from the T-DNA region of the Ti plasmid would be potentially useful for this purpose. In chloroplasts, however, the expression of *nif* may not require such an extensive degree of manipulation, since chloroplast gene expression appears to be compatible with that of prokaryotic organisms. It may be that chloroplast RNA polymerase could initiate transcription at the *nif* promoters in the presence of *nif* activator protein.

We thank Mike Merrick for constructive criticism of this manuscript.

REFERENCES (Postgate & Cannon)

Ausubel, F. M., Bird, S. C., Durbin, K. J., Janssen, K. A., Margolskee, R. F. & Perkin, A. P. 1979 Glutamine synthetase mutations which affect expression of nitrogen fixation genes in *Klebsiella pneumoniae*. *J. Bact.* **140**, 597–606.

Ausubel, F. M., Margolskee, R. & Maizels, N. 1977 Mutants of *Klebsiella pneumoniae* in which expression of nitrogenase is independent of glutamine synthetase control. In *Recent developments in nitrogen fixation* (ed. W. Newton, J. R. Postgate & C. Rodriguez-Barrueco), pp. 347–356. New York: Academic Press.

Barber, L. E., Tjepkema, T. P. & Evans, H. J. 1978 Acetylene reduction in the root environment of soybeans, grasses and other plants in Oregon. *Ecol. Bull., Stockh.* **26**, 366–372.

Cannon, F. C. & Postgate, J. R. 1976 Expression of nitrogen fixation genes *nif* in *Azotobacter*. *Nature, Lond.* **260**, 271–272.

Cannon, F. C., Dixon, R. A. & Postgate, J. R. 1976 Derivation and properties of F-prime factors carrying nitrogen-fixation genes from *Klebsiella pneumoniae*. *J. gen. Microbiol.* **93**, 111–125.

Cannon, F. C., Riedel, G. E. & Ausubel, F. M. 1977 A recombinant plasmid which carries part of the nitrogen fixation (*nif*) gene cluster of *Klebsiella pneumoniae*. *Proc. natn. Acad. Sci. U.S.A.* **74**, 2963–2967.

Cannon, F. C., Riedel, G. E. & Ausubel, F. M. 1979 Overlapping sequences of *Klebsiella pneumoniae nif* DNA cloned and characterised. *Molec. gen. Genet.* **174**, 59–66.

Collmer, A. & Lamborg, M. 1976 Arrangement and regulation of nitrogen fixation genes in *Klebsiella pneumoniae* studied by derepression kinetics. *J. Bact.* **126**, 806–813.

Dixon, R. A. & Postgate, J. R. 1972 Genetic transfer of nitrogen fixation from *Klebsiella pneumoniae* to *Escherichia coli*. *Nature, Lond.* **237**, 102–103.

Dixon, R. A., Cannon, F. C. & Kondorosi, A. 1976 Construction of a P plasmid carrying nitrogen fixation genes from *Klebsiella pneumoniae*. *Nature, Lond.* **260**, 268–271.

Dixon, R. A., Eady, R., Espin, G., Hill, S., Iaccarino, M., Kahn, D. & Merrick, M. 1980 Analysis of the regulation of the *Klebsiella pneumoniae* nitrogen fixation (*nif*) gene cluster with gene fusions. *Nature, Lond.* **286**, 128–132.

Dixon, R. A., Kennedy, C., Kondorosi, A., Krishnapillai, V. & Merrick, M. 1977 Complementation analysis of *Klebsiella pneumoniae* mutants defective in nitrogen fixation. *Molec. gen. Genet.* **157**, 189–198.

Dobereiner, J. & Day, J. M. 1976 Associative symbioses in tropical grasses: characterization of microorganisms and dinitrogen-fixing sites. In *Proceedings of the 1st International Symposium on Nitrogen Fixation* (ed. W. E. Newton & C. J. Nyman), vol. 2, pp. 518–538. Pullman: Washington State University Press.

Dobereiner, J. & De-Polli, H. 1980 Diazotrophic rhizocoenoses. In *Nitrogen fixation* (ed. W. D. P. Stewart & J. R. Gallon) (*Proc. Phytochem. Soc. Eur.* no. 18), pp. 301–333. London: Academic Press.

Dobereiner, J., Day, J. M. & Dart, P. J. 1972 Nitrogenase activity and oxygen sensitivity of the *Paspalum notatum-Azotobacter paspali* association. *J. gen. Microbiol.* **71**, 103–116.

Eady, R. R., Issack, R., Kennedy, C., Postgate, J. R. & Ratcliffe, H. 1978 Nitrogenase synthesis in *Klebsiella pneumoniae*: comparison of ammonium and oxygen regulation. *J. gen. Microbiol.* **104**, 277–285.

Elmerich, C., Tandeau de Marsac, N., Chocat, P., Charfin, N., Aubert, J. P., Gerbaud, C. & Guerineau, M. 1980 Cloning of the nitrogen fixation (*nif*) genes in *Klebsiella pneumoniae* in a chimeric cosmid and transformation of yeast *Saccharomyces cerevisiae*. In *Proceedings of the 5th European Meeting on Bacterial Transformation and Transfection*, Florence, 2–5 September. (In the press.)

Evans, H. J., Emerich, D. W., Ruiz-Argueso, T., Maier, R. J. & Albrecht, S. L. 1980 Hydrogen metabolism in the legume–Rhizobium symbosis. In *Nitrogen fixation*, vol. 2 (ed. W. E. Newton & W. H. Orme-Johnson), pp. 69–86. Baltimore: University Park Press.

Fink, G. R. 1980 Unusual genetic events associated with a transposable element in yeast. In *Proceedings of the 5th European Meeting on Bacterial Transformation and Transfection*, Florence, 2–5 September. (In the press.)

Giles, K. L. & Whitehead, H. C. M. 1975 The transfer of nitrogen-fixing ability to a eukaryotic cell. *Cytobios* **14**, 49–61.

Hardy, R. W. F. 1976 Potential impact of current abiological and biological research on the problem of providing fixed nitrogen. In *Proceedings of the 1st International Symposium on Nitrogen Fixation* (ed. W. E. Newton & C. J. Nyman), vol. 2, pp. 693–717. Pullman: Washington State University Press.

Hardy, R. W. F. & Havelka, U. D. 1976 Photosynthate as a major factor limiting nitrogen fixation by field-grown legumes in the emphasis on soybeans. In *Symbiotic nitrogen fixation in plants* (ed. P. S. Nutman), pp. 421–428. Cambridge University Press.

Hernalsteens, J. P., Van Vliet, F., De Beuckeleer, M., Depicker, A., Engler, G., Lemmers, M., Holsters, M., Van Montagu, M. & Schell, J. 1980 The *Agrobacterium tumefaciens* Ti plasmid as a host vector system for introducing foreign DNA in plant cells. *Nature, Lond.* **287**, 654–656.

Hill, S. & Kavanagh, E. 1980 Roles of *nif*F and *nif*J gene products in electron transport to nitrogenase in *Klebsiella pneumoniae*. *J. Bact.* **141**, 470–475.

Hohn, B., Lebeurier, G. & Hohn, T. 1980 Cloning with cosmids in pro- and eukaryotes. In *Proceedings of the 13th FEBS Meeting*, Jerusalem, 24–29 August, p. 233.

Janssen, K. A., Riedel, G. E., Ausubel, F. M. & Cannon, F. C. 1980 Transcriptional studies with cloned nitrogen fixation genes. In *Nitrogen fixation*, vol. 1 (ed. W. E. Newton & W. H. Orme-Johnson), pp. 85–93. Baltimore: University Park Press.

Krishnapillai, V. & Postgate, J. R. 1980 Expression of Klebsiella *his* and *nif* genes in *Serratia marcescens*, *Erwinia herbicola* and *Proteus mirabilis*. *Arch. Microbiol.* **127**, 115–118.

Kustu, S., Burton, D., Garcia, E., McCarter, L. & McFarland, N. 1979 Nitrogen control in *Salmonella*: regulation by the *gln*R and *gln*F gene products. *Proc. natn. Acad. Sci. U.S.A.* **76**, 4576–4580.

Larson, R. I. & Neal, J. L. 1978 Selective colonization of the rhizosphere of wheat by nitrogen-fixing bacteria. *Ecol. Bull., Stockh.* **26**, 331–342.

MacNeil, D. & Brill, W. J. 1980 Isolation and characterization of λ specialized transducing bacteriophages carrying *Klebsiella pneumoniae nif* genes. *J. Bact.* **141**, 1264–1271.

MacNeil, T., MacNeil, D., Roberts, G. P., Supiano, M. A. & Brill, W. J. 1978 Fine-structure mapping and complementation analysis of *nif* (nitrogen fixation) genes in *Klebsiella pneumoniae*. *J. Bact.* **136**, 253–266.

Merrick, M., Filser, M., Dixon, R., Elmerich, C., Sibold, L. & Houmard, J. 1980 Use of translocatable genetic elements to construct a fine-structure map of the *Klebsiella pneumoniae* nitrogen fixation (*nif*) gene cluster. *J. gen. Microbiol.* **117**, 509–520.

Mortenson, L. E. & Thorneley, R. N. F. 1979 Structure and function of nitrogenase. *A. Rev. Biochem.* **48**, 387–418.

Nieva-Gomez, D., Roberts, G. P., Klevickis, S. & Brill, W. J. 1980 Electron transport to nitrogenase in *Klebsiella pneumoniae*. *Proc. natn. Acad. Sci. U.S.A.* **77**, 2555–2558.

Pahel, G. & Tyler, B. 1979 A new *gln*A-linked regulatory gene for glutamine synthetase in *Escherichia coli*. *Proc. natn. Acad. Sci. U.S.A.* **76**, 4544–4548.

Parejko, R. A. & Wilson, P. W. 1970 Regulation of nitrogenase synthesis by *Klebsiella pneumoniae*. *Can. J. Microbiol.* **16**, 681–685.

Postgate, J. R. 1980a The nitrogen economy of marine and land environments. In *Food chains and human nutrition* (ed. Sir Kenneth Blaxter), pp. 161–185. London: Applied Science Publishers.

Postgate, J. R. 1980b Prospects for the exploitation of biological nitrogen fixation. *Phil. Trans. R. Soc. Lond.* B **290**, 421–425.

Postgate, J. R. & Krishnapillai, V. 1977 Expression of Klebsiella *nif* and *his* genes in *Salmonella typhimurium*. *J. gen. Microbiol.* **98**, 379–385.

Puhler, A. & Klipp, W. 1981 Fine structure analysis of the gene region for N_2-fixation (*nif*) of *Klebsiella pneumoniae*. In *Biological metabolism of inorganic nitrogen and sulfur compounds* (ed. H. Bothe & A. Trebst). Berlin, Heidelberg and New York: Springer-Verlag. (In the press.)

Puhler, A., Burkhardt, H. J. & Klipp, W. 1979a Cloning of the entire region for nitrogen fixation from *Klebsiella pneumoniae* in a multicopy plasmid vehicle in *Escherichia coli*. *Molec. gen. Genet.* **176**, 17–24.

Puhler, A., Burkhardt, H. J. & Klipp, W. 1979b Cloning in *Escherichia coli* the genomic region of *Klebsiella pneumoniae* which encodes genes responsible for nitrogen fixation. In *Plasmids of medical, environmental and commercial importance* (ed. K. N. Timmis & A. Puhler), vol. 1, pp. 435–441.

Riedel, G. E., Ausubel, F. M. & Cannon, F. C. 1979 Physical map of chromosomal nitrogen fixation (*nif*) genes of *Klebsiella pneumoniae*. *Proc. natn. Acad. Sci. U.S.A.* **76**, 2866–2870.

Roberts, G. P., MacNeil, T., MacNeil, D. & Brill, W. J. 1978 Regulation and characterisation of protein products coded by the *nif* (nitrogen fixation) genes of *Klebsiella pneumoniae*. *J. Bact.* **136**, 267–279.

Ruiz-Argueso, T., Hanus, J. & Evans, H. J. 1978 Hydrogen production and uptake by pea nodules as affected by strains of *Rhizobium leguminosarum*. *Arch. Microbiol.* **116**, 113–118.

St John, R. T., Shah, V. K. & Brill, W. J. 1974 Regulation of nitrogenase synthesis by oxygen in *Klebsiella pneumoniae*. *J. Bact.* **119**, 266–272.

Sprent, J. I. 1979 *The biology of nitrogen-fixing organisms*. London: McGraw-Hill.

Streicher, S. L., Shanmugam, K. T., Ausubel, F., Morandi, C. & Goldberg, R. B. 1974 Regulation of nitrogen fixation in *Klebsiella pneumoniae*: evidence for a role of glutamine synthetase as a regulator of nitrogenase synthesis. *J. Bact.* **120**, 815–821.

Tubb, R. S. 1974 Glutamine synthetase and ammonium regulation of nitrogenase synthesis in *Klebsiella*. *Nature, Lond.* **251**, 481–485.

Tubb, R. S. & Postgate, J. R. 1973 Control of nitrogenase synthesis in *Klebsiella pneumoniae*. *J. gen. Microbiol.* **79**, 103–117.

Tyler, B. 1978 Regulation of the assimilation of nitrogen compounds. *A. Rev. Biochem.* **47**, 1127–1162.

Van Ardel, J. G. & Brown, C. M. 1977 Ammonia assimilation in the fission yeast *Schizosaccharomyces pombe* 972. *Arch. Microbiol.* **111**, 265–270.

Yamada, T. & Sakaguchi, K. 1980 Nitrogen fixation associated with a hotspring green alga. *Arch. Microbiol.* **124**, 161–167.

Discussion

G. PONTECORVO, F.R.S. Would the energy consumption of nitrogen fixation not be deleterious in any new diazotrophic system?

J. R. POSTGATE. I thank Professor Pontecorvo for asking that question, because already in private discussion I have met several people who regard the ATP demand of nitrogen fixation as 'enormous' or 'crippling'. This is not true, so I think the point needs clarifying.

One can write a general equation for biological nitrogen fixation so:

$$N_2 + 3NADH + 5H^+ + 12ATP \longrightarrow 2NH_4^+ + 3NAD + 12ADP.$$

The 12 ATPs consumed by nitrogenase are the source of the oft-quoted 'huge energy demand' of nitrogenase because, on thermodynamic grounds, one would expect little or no energy loss in the overall reaction. To those 12 ATPs one must add 9 ATPs that could have been generated from 3 NADH (at a P/O ratio of 3) had they not been deflected into diazotrophy. Thus the basic cost of nitrogenase function becomes 10.5 ATP lost per N atom brought to the level of NH_4^+. I shall mention factors that alter that estimate later.

This is a crude estimate of the energy debt to an organism if it uses N_2 instead of NH_4^+. However, plants very rarely use NH_4^+; they use nitrate. One can write a corresponding general equation for nitrate reduction so:

$$NO_3^- + 4NADH + 6H^+ \longrightarrow NH_4^+ + 4NAD + 3H_2O.$$

The consumption of 4 NADH brings the basic cost of nitrate reduction to 12 ATP lost per N atom brought to the level of NH_4^+.

The superficial impression that nitrate reduction is less efficient than diazotrophy is something of an illusion because these calculations exclude: (i) loss of H_2 by a side reaction of nitrogenase, which can add another ATP; (ii) ATP consumption in assimilation of NH_3, which is probably much the same for both processes; (iii) the energy cost of transporting NO_3^- into the cells, which diazotrophy probably avoids; (iv) the fact that 12 ATP/N_2 is a minimum figure for consumption by nitrogen as nitrogenase; (v) the energy costs of synthesizing and maintaining nitrogenase compared with nitrate reductase; and (vi) the differential effects of the localization of the two processes in plants (e.g. in roots compared with leaves). Some of these factors will alter the energy budget substantially in practice, the trend being to favour nitrate reduction over diazotrophy, but the important message is that, as *biochemical processes*, diazotrophy and nitrate reduction can be expected to make very similar demands on the plant's energy budget. This is indeed borne out in plant experiments (see, for example, Minchin & Pate 1973; Silsbury 1977), where the differences between nitrate-grown and N_2-grown legumes can be quite small.

References

Minchin, F. R. & Pate, J. S. 1973 The carbon balance of a legume and the functional economy of its root nodules. *J. exp. Bot.* **24**, 259–271.

Silsbury, J. H. 1977 Energy requirement for symbiotic nitrogen fixation. *Nature, Lond.* **267**, 149–150.

Perspective and prospect

BY SIR KENNETH MATHER, F.R.S.

Department of Genetics, University of Birmingham, Birmingham B15 2TT, U.K.

INTRODUCTION

Plant breeding is concerned with the production of improved varieties of plants by the development of superior genotypes. The nature of the improvements that are sought will not in general depend on genetical considerations, though in some circumstances it may be influenced by them. Rather it will depend on such things as the requirements, preferences or even idiosyncrasies of the consumer, on economic considerations and on the agronomic needs and practices of the grower. Alternative ways of meeting these requirements must be considered alongside the approach through plant breeding. All these must be taken into account by the plant breeder in setting his targets, which must pay due regard to feasibility and also have a sufficient element of prediction, or prophecy, in them to allow for the necessary lapse of time between setting up the breeding programme and having the finished variety ready for the market.

Having set his targets, the breeder must ask how he can best proceed towards achieving them; which features of the plant's physiology offer him the best prospect of raising its production under the various circumstances in which it is likely to be grown; how he can secure the best distribution of assimilates to the economically important part of the plant as opposed to parts not so economically utilizable; how losses arising from the depredation of pests and diseases can be minimized, and so on. The targets must, in fact, be expressed in terms of more detailed definable characters that the breeder will seek to adjust, and which can be followed relatively easily in single individuals, or small families, as a basis for effective comparison and hence meaningful selection. In this the breeder must seek the cooperation of physiologist and pathologist, and also, because of changing practices of agriculture, of agronomist and agricultural engineer.

The part that the physiologist can play, and is indeed already coming to play, in plant breeding emerges clearly from the papers presented by Professor Cooper and Mr Bingham. The latter also refers to the problems of breeding for resistance to pests and diseases, and particularly to the transience so frequently met in the disease resistance with which the breeder endows his varieties. The ability to recognize durable disease resistance, other than by the belated test of its survival in practice, is surely one of the greatest needs of plant breeding today, and one that can be met only with the cooperation of pathologists. Engineering problems raised, for example, by the mechanical harvesting of root and fruit crops are not mentioned; but here, too, breeder and engineer must work together to overcome them.

Plant breeding and genetics

The technology of plant breeding thus transcends genetics when setting its targets, assessing their feasibility, and defining the characters with which the breeder must concern himself. At the same time, since the breeder is seeking to achieve his ends by adjustment of the genotype, his approach must be through applied genetics, and any advance in genetical science will thus be of interest to him as prospectively offering means of refining or extending the methodology on which he can draw for his purposes. The question has, however, frequently been asked, and is indeed still asked on occasion, 'What *has* genetics contributed to plant breeding?' To begin answering this question we must go back to the turn of this century.

Although hybridization was not unknown to the breeders of production crops in the nineteenth century, it appears to have played little part in their work. Their emphasis was on the method of selection, and even in the 1890s the dramatic improvements made by Hjälmar Nilsson in the wheat, barley and oats of Sweden were achieved through single-ear selections from whatever variants presented themselves in the materials available to him. There seems to have been little appreciation of hybridization as a means of deliberately bringing together in one variety desirable features from two or more parental lines. All this had changed by 1906 when, as Dr G. D. H. Bell, F.R.S. has kindly drawn to my attention, R. H. Biffen wrote, 'As recently as six years ago the chances of . . . (obtaining the type of barley we desire by crossbreeding) . . . were so small that the attempt was barely worth the making. Since then thanks to the discoveries of Gregor Mendel . . . we can with reasonable certainty of obtaining the results we require begin our labours. The whole aspect of plant-breeding has been changed and where formerly all was chaos we can now perceive order and the action of definite laws. A few years ago cross-breeding was a gamble of the wildest description, now we can calculate the result of combining any two given parents with well-nigh mathematical accuracy.' He also wrote, 'I can foresee the time coming when we shall be able to unite in one variety the best quality, the best cropping power, the best straw and so on, which we can find in all the numerous varieties now in existence.' Such was the impact of Mendel's findings when, with the principles of segregation and recombination, they gave plant breeders a firm foundation for their practice. Techniques of selection were still important, but now the breeder could with confidence seek to construct the superior genotype for his selection to pick out. And as such, this development can fairly be regarded as the beginning of the manipulation of genetic systems in plant breeding.

This deliberate construction of desirable genotypes is illustrated in a different way by the revolutionary development of hybrid corn, which came along some years later in the U.S.A. Can we doubt that this, too, with its successful combination of controlled uniformity and the heterosis characteristic of hybrids in this crop, stemmed from an appreciation of Mendel's principles, fortified perhaps by Johannsen's experimental demonstration of pure lines and their properties? Hybridization to bring together, after segregation and recombination, desirable features from several parental lines into a single variety or line, and hybridization of chosen inbred lines to give F_1s for immediate use in production are the foundation of our present methods of plant breeding, which Professor Williams describes and discusses.

While Mendel's principles were, of course, his most important legacy to us, he also gave a precise prescription for bringing together (or separating) genes whose individual effects on the phenotype could be recognized without ambiguity. This led to high hopes, no doubt made all the higher by findings such as that of Biffen when he showed in 1907 that rust resistance in wheat could segregate from susceptibility in the manner of a single gene difference – hopes that plant breeding would be reduced to little more than an exercise in Mendelian manipulation. Disappointment was thus all the greater when these hopes foundered on the rock of quantitative variation, which appeared not be to capable of Mendelian analysis or even at first of Mendelian interpretation: most of the variation that the plant breeder has to deal with is of this kind. It has taken a long time to show unequivocally that heritable quantitative variation is as dependent on nuclear genes as the simpler Mendelian differences are; to erect on this foundation a methodology of biometrical genetics that enables us to analyse the variation into its non-heritable and various heritable components; and to set out their implications for the work of the plant-breeder. Professor Jinks, however, tells us of the power that biometrical genetics has now come to achieve in tracing the causation of such basic phenomena as heterosis, and in predicting the outcome of breeding programmes. It can thus help the breeder not only to interpret results that he obtains, but also to plan the strategies of his breeding programmes; and it is all the more useful for these purposes in that it deals with the totality of variation in any character with which he is concerned.

Chromosomes, genomes and recombination

Biometrical genetics does not depend for its analytical capacity on assigning to individual chromosomes the genes that contribute to quantitative variation or on assessing the contributions that individual chromosomes make to the totality of the variation observed: indeed, the methodology is of general use and is applicable to all species just because of its capacity for dealing with the totality of variation in a population or a cross, without any such initial partitioning. At the same time, the value of being able to assess the contributions of individual chromosomes can be of great value in facilitating not only the analysis, but even more the synthesis, of desirable combinations of genes, as has long been apparent from studies with *Drosophila*. Facilities for assaying the individual chromosomes were unique to *Drosophila* until relatively recently. Now, however, they are available also in wheat, as Dr Law and Professor Driscoll emphasize to us. We can ascertain how the variation of a character in wheat is distributed between chromosomes and between genotypes; the chromosomes can be deliberately put together to give superior genotypes; valuable blocks of genes can be moved from one variety to another; and special types, such as male-steriles, can be made where they are needed. The technique of manipulating the chromosomes is not the same as in *Drosophila* in that it depends on following and managing them by cytological rather than genetical means; but it is equally effective. At present its use is confined in the main to wheat and its relatives, though, as Dr Thomas tells us, it is becoming increasingly possible in oats, with correspondingly valuable results. Chromosome manipulation, as we see it in these cereals, is indeed a powerful tool for the breeder wherever it can be employed, though its application will necessarily be restricted to species that have chromosomes capable of being followed cytologically as individuals, and whose constitution is such that they can tolerate any genic unbalance consequent on aneuploidy.

The chromosome analysis of wheat has revealed the special action of a gene, or genes, borne on chromosome 5B, in restricting pairing and hence recombination to homologues: in the absence of this chromosome, or when its effect is suppressed, pairing and recombination takes place between homoeologues as well as between homologues. Dr Riley (with Dr Law and Dr Chapman) describes how this can be used to incorporate into bread wheat, carrying the normal 42 chromosomes, desirable characters, such as rust resistance and baking quality, from *Aegilops* spp. In emphasizing the importance of recombination for the management of the variation upon which selection is practised, he notes that at times the breeder's requirements make it necessary to aim for reduction of recombination but at other times for its increase, and he points out that, as a result of the cytogeneticist's work, the breeder now has the means of intervening, to his own benefit, in the genetic control of recombination.

Turning to a different type of manipulation, of chromosome sets rather than of individual chromosomes, the ability to provide haploid individuals when wanted can be a powerful aid to the assessment of quantitative variation, and also to the identification and fixation of desirable recombinants. Haploidy can arise, and indeed be induced, in a variety of ways that are listed and discussed by Dr Hermsen (with Dr Ramanna). It may arise after polyembryony or as a result of pseudogamy. It can be induced in barley by crossing with other species of *Hordeum*, the chromosomes of the alien species being eliminated during embryo formation, and it can be induced elsewhere by pollen culture. No matter how they are obtained, the haploids can easily be rediploidized whenever this is desired and they will then give fully homozygous lines for use whether as varieties in their own right or as 'inbred' parents for intercrossing in a hybrid breeding programme. Nor is the value of the technique restricted to its use in diploid species: it can equally be used for at least the partial analysis, and subsequent resynthesis, of polyploids like the potato.

Polyploids

Chromosome studies early showed that many of our crop plants, including some of the most important, are polyploids, most commonly allopolyploids (or amphidiploids if the term is preferred), though the grass *Dactylis glomerata* may be an autotetraploid. With the discovery of the capacity of colchicine for doubling the number of chromosomes, it became possible to make polyploids at will, whether autopolyploids by the direct doubling of existing species or allopolyploids by doubling hybrids between species capable of being crossed. High hopes were entertained of induced polyploids, especially allopolyploids, as a means of plant improvement, but only limited success has been obtained so far. Autopolyploidy has proved of some interest with ornamental plants in which the value of the increased size that it often engenders can more than offset the reduction in fertility that also results. This could make it of some value in forage crops, too.

In allopolyploidy, the combination of different genomes is likely to bring about greater adaptability rather than, of itself, to produce increase in yield or quality, as Dr Breese (with Dr Lewis and Dr Evans) points out. Furthermore, bringing together sets of genes that have not been adjusted by natural selection to work harmoniously with one another may well result in genic unbalance and hence a certain amount of trouble, which must be removed by selective breeding. For these various reasons a newly synthesized polyploid grain crop is likely to require considerable attention from the breeder before it becomes commercially utilizable, and experience with triticale confirms this expectation. Herbage grasses, as Dr Breese further observes, appear to offer the greatest prospective rewards from hybrid polyploids, for flexibility is a prime requirement in swards while the harvest depends on the vegetative growth rather than seed production (though, of course, enough seed must be obtainable for the commercial propagation of the polyploid). Dr Breese and his collaborators have in fact produced agronomically successful tetraploid hybrids between the two diploid grass species *Lolium perenne* and *L. multiflorum*, and though not autotetraploid these hybrids are not full allopolyploids either: some pairing occurs between homoeologues, although it is preferentially between homologues. This preference in pairing is subject to genic control and may be enhanced by selection, so raising the possibility of moving towards the genetic stability of a true allopolyploid. B chromosomes are also known to increase preferential pairing in other tetraploid hybrids between *Lolium* species, so offering a further means of adjusting them to the breeder's requirements.

Other allopolyploids between species of *Lolium* and *Festuca* may well prove to have more value for the transfer of desirable genes and gene complexes from one parent species to the other than as production crops in their own right. It is clear, however, that artificial polyploids have at last come to stay, at least in grass breeding, and basic problems of gene action, particularly in relation to quantitative variation, would now well repay study in them.

New variants

Heritable variation is the plant breeder's raw material, and if it is inadequate for his purposes and is not readily available in other ways, he may seek to induce it by means of radiation or chemical mutagens. Professor Nilan reviews the use of induced changes of both genes and chromosomes in crop plant breeding. There can be no doubt that such gene mutations can be of great value in the investigation of the biochemistry and physiology of important characters and of the genetic structure underlying their manifestion. Neither can there be any doubt of the value, or prospective value, of induced gene mutation to the plant breeder under special circumstances. They have been the basis of the successful development of increasingly high levels of antibiotic production by the fungi on which the antibiotic industry is based: indeed with imperfect fungi their use is unavoidable. They also offer a prospectively valuable approach to, for example, the production of new forms of the standard varieties of top fruit, such as the apple, differing from the parent form in, say, having a size or structure of tree that facilitates harvesting or in being self-fertile, but yielding fruit that is still of the same kind as the parent and so avoids the hazard inevitably met by a grower who seeks to introduce, necessarily at considerable expense, a recognizably new variety in a field where consumer preference is both strong and conservative. As a general adjunct to the breeding of regular crop plants, however, its value is less clear to me. Indeed, mutations of major effect appear seldom to be of a kind that the breeder would regard as prospective raw material not obtainable more usefully in other ways. Furthermore, if we may judge by findings from *Drosophila*, induced genic mutations are not likely to be a very useful means of augmenting the selectable quantitative variation upon which depends so much of the breeder's progress.

Mutagens also, of course, bring about structurally changed chromosomes. As Professor Nilan points out, these can be of great value in the genetic analysis of crop plants and in chromosome manipulation, as for example in Professor Driscoll's production of male-steriles in wheat.

During the course of this symposium, reference has been made to a further, and prospectively most valuable, technique for manipulating genetic variation by the use of ionizing radiation. With his collaborators, Professor Jinks (who, however, does not himself describe the technique) has been pollinating flowers of *Nicotiana rustica* with pollen that has been subjected to a near-lethal dose of radiation, and which came from a different line of this species. When germinated, the seed obtained in small quantities from these pollinations produced plants just like their mothers in nearly all respects, but showing evidence of having derived single marker genes from their fathers (Virk et al. 1977). There was evidence too, in other cases, of genes affecting quantitative variation having been transmitted in the same way. Given that these paternal genes become stably incorporated into the genotype of the progeny plants, as results obtained by Pandey (1975, 1978) suggest can happen, this technique would offer a way of transferring desirable genes from one line to another with little disturbance of the rest of the genotype and much more quickly than can be achieved by the customary repeated backcrossing. Further results from the use of irradiated pollen from a different species suggest that this approach may also offer the very valuable possibility of transferring desirable genes in the same way between species that do not hybridize effectively by normal pollination.

The value of mutagens lies in their ability to increase the frequency of changes in the nuclear materials. It has, however, long been known that abnormalities of nuclear behaviour can arise as a result of both gene mutation and general upset of genic balance (see Darlington 1932, 1939; Rees 1961). The regular behaviour of nucleus and chromosomes that we normally see must thus have been achieved and maintained by adjustment of the genotype through natural selection. Even so, upsets do occasionally occur, as for example when normally diploid plants produce a tetraploid sector, which is, however, usually contained and limited by competition from surrounding cells of the diploid constitution. In discussing the significance for the plant breeder of spontaneous nuclear upsets, Dr Bennett points to both the nuisance value of their consequences for the maintenance of standards of genetical purity in commercial crops, and the possibilities that they open up for the breeder. He cites three examples of instability that can have direct consequences for the breeder. One is the result of a gene mutation, *tri*, in barley, and the others are the outcome of the genic unbalance consequent on wide outcrossing, namely selective chromosome elimination in *Hordeum* species crosses and the case of the amphidiploid triticale. It is of special interest that, as Dr Bennett tells us, improvement in the quality of seed in triticale has been achieved by a cytogenetic approach.

Dr Durrant is concerned with genic rather than chromosomal instability. In the light of the known instabilities of major genes in *Antirrhinum* and other plants he discusses the possibility of similar instabilities in the genes mediating quantitative variation, where their consequences would be virtually impossible to distinguish from variation arising from other more readily recognizable causes such as segregation or the impact of the environment. He considers whether the remarkable phenomenon of 'conditioning', which he discovered in flax and which has since been observed in *Nicotiana rustica*, could be the expression of environmentally induced instabilities of this kind. We still have not come to a full understanding of this tantalizing phenomenon, though our information about it continues to increase. It is currently known only in these two species of plant, but a fuller understanding might show how it could be traced elsewhere and possibly put to good use in breeding production crops.

Tissues, cells and protoplasts

The need for sexual propagation by seed, or even clonal propagation by special organs such as tubers, necessarily imposes constraints on the isolation and exploitation of new variants, especially somatic variants, as well as on their induction. It limits, too, even the range over which wide hybridization can be attempted. To remove, or at least ease, these constraints could greatly facilitate the manipulation of genetic systems, and it could also be of help to the plant breeder in other ways.

Professor Davies tells us how this is now being attempted by cell and tissue culture. In doing so he cites a dozen different ways in which the plant breeder can be served by these techniques, ranging from the storage and rapid multiplication of valuable genotypes in a disease-free condition, to the isolation of somatic variants that could otherwise be lost through cell competition and including, for example, the investigation of conditioning. He quotes examples where this approach has already met with success, and we may note also the remarkable series of variants, many of prospective practical value, obtained by Shepard et al. (1980) from the cloning of isolated protoplasts of the potato Russet Burbank, whose use in normal breeding programmes is limited by its sterility.

There are, of course, limitations to this approach. Some are analytical in that, of itself, the isolation and propagation of variants by cell and tissue culture, leading to the regeneration of whole variant plants, cannot regularly tell us the causal nature of the variation, whether nuclear or extranuclear, and whether capable or incapable of transmission through seed. This difficulty could, however, generally be overcome by the direct test of breeding from the regenerated plants. Other limitations are technical, including the need for being able to regenerate complete plants from the tissues or cells. At present this can be done for only a restricted range of species, but the problems will no doubt be overcome in time by experimental persistence and growing physiological know-how in species that are of importance to the breeder.

Professor Cocking confines his discussion to the prospective value of protoplasts in genetic manipulation. While noting the successful use of protoplasts for recovering variants in the potato (already mentioned above), he points especially to two other potential uses of protoplasts. One is in the production of somatic hybrids obtained by fusion of protoplasts from the two parental lines, so freeing hybridization from constraints imposed by the sexual system. Indeed, somatic hybrids have already been made between species that will not cross sexually, as well as between others that will. The second potential use that he stresses is in the transfer of genes between species, whether by a technique parallel to that with the use of heavily irradiated pollen or by transformation by using *Agrobacterium* plasmids. It is not yet known whether hybrid polyploids made somatically would have properties different from those of their sexually produced counterparts, but there is no compelling reason *a priori* why they should. Transformation by infection with suitably engineered plasmids, carrying for example the *nif* system of genes, may, however, have wide possibilities of great importance.

The use of protoplasts, like that of cells, is, of course, constrained by the need to regenerate whole plants from them, but the recent regeneration, achieved with the use of young immature leaf material in *Sorghum* (Wernicke & Brettell 1980) encourages the belief that it will soon also be achieved in our cereals.

The molecular approach

The rise of molecular genetics has given us a growing understanding of DNA in all its variety of forms, as well as a growing range of techniques for both analysing and reorganizing it. In doing so it has opened new and wider, even if more distant, prospects for the manipulation of the genetic materials in the amelioration of our plants for the purposes of production. Nor are these advances confined to the DNA of the chromosomes: that of chloroplasts and mitochondria come into the picture, too, with a further prospective widening of the understanding of the materials that the plant breeder, implicitly or explicitly, is using; of the problems that he may have to face in using them; and of the ways in which he may hope to exploit them. Three of the papers reflect this molecular approach, each in its own way.

Professor Rees (with Dr Narayan) is concerned with what he aptly describes as the astonishing variation in the amounts of chromosome DNA in species of higher plants. Much of the variation is due to amplification of sequences: the proportion of repetitive DNA rises disproportionately as the total increases, in some cases to as much as 70%. The function of this repetitive DNA is unclear. It does not affect the viability of, for example, gametes; but it does affect the distribution of chiasmata in the chromosomes and hence the pattern of recombination. It carries few genes coding for proteins, but it appears to contribute to quantitative variation. Can its manipulation be put to use in plant breeding?

Dr Flavell has been cloning DNA of various kinds and various origins. Such cloning, as he points out, makes possible the physical mapping of chromosome material, whether from the nucleus or elsewhere, irrespective of whether or not it is the seat of variation detectable by normal genetic methods. It has already permitted a molecular analysis of the telomeric heterochromatin in rye, whose adjustment, as we have already seen, produces a reduction in the incidence of shrivelled grain in triticale. And the cloning of specific sequences from the mitochondrial DNA of maize has provided means of recognizing cytoplasms associated with male sterility even where the presence of restorer genes prevents manifestation of the sterility. One of the cytoplasms now capable of recognition in this way has been widely used in hybrid corn breeding and has, as a result, led to a disastrously widespread loss of resistance to Southern corn blight, a fungal disease. Dr Flavell finally emphasizes the possibilities that cloning opens up for the modification of plant genotypes by the insertion of specific genes into them.

In a *tour de force* of coverage, clarity and realism, Professor Postgate (with Dr Cannon) turns to the biological fixation of nitrogen in relation to green plants. It might be raised by breeding more effective strains of *Rhizobium*, and other commensal microorganisms might have *nif* genes inserted into them. But the really fundamental approach is through the introduction of the *nif* complex, which includes some 17 genes, into the green plants themselves and securing their expression there. This raises formidable problems not only in relation to the physiology and action of the genes themselves, but also of securing their entry into the higher plant cells by means of a vector that would either maintain a satisfactory numerical stability in the somatic cells of the higher plant and also pass through seed, or preferably would bring about the incorporation of the *nif* complex into the plant's chromosomes. Initial entry would presumably be into protoplasts and the need for regenerating whole plants from them would again arise. Professor Postgate does not disguise the magnitude of these problems and he estimates that it will take a decade or more before we can know whether such engineering is possible. At present, however, he sees no reason to doubt its feasibility.

Conclusion

After the establishment of Mendel's basic principles the notion of deliberately producing superior genotypes by bringing together desirable genes through hybridization, segregation and recombination, took on vigorous life in the minds of plant breeders. Such has been its success and so widespread is it that it is now regarded as conventional plant breeding. It has had its difficulties, notably in the utilization and management of quantitative variation, but this is now yielding to new and appropriate methods of genetic analysis.

Later developments of genetical science are coming to have their impact too. The rise of cytogenetics, in which Professor Darlington, who introduces this symposium, played such a fundamental part, has led to chromosome manipulation, and this is already achieving its successes. The induction of changes in genes and chromosomes by the use of ionizing radiations and other mutagenic agencies is offering us a variety of means for producing genetic changes where we need them, for facilitating chromosome manipulation and for transferring particularly desirable genes expeditiously from one form, or even species, into another. Now the techniques of tissue, cell and protoplast culture, and, still more fundamentally, those of molecular genetics, are opening up yet further prospects and promising yet further rewards. Beyond even these we can now begin to discern, however dimly as yet, the possibility of deliberately manufacturing genes themselves, with properties that we can specify in advance.

Some new techniques will fall by the wayside, and all will raise their problems which must be overcome. We can, however, see a progression of methodologies from the conventional, which are now accepted as normal practice, through the feasible, which are beginning to record their successes, to the notional, where the problems are still awaiting solution. But can we doubt that the phenominal and accelerating progress of genetical science, supported by that in cognate disciplines, will turn the notional of today into the feasible of tomorrow and the conventional of next week or the week after?

Some of the advances, achieved and prospective, will be of general application; some, however, will be restricted to particular kinds of crop, to particular kinds of genotype or even to particular species, according to the needs that these reveal or the opportunities that they offer. Some advances will depend on developments in physiology or pathology as much as genetics. But all will, in their own ways and their own times, come to open up new prospects for the plant breeder. At the same time, such is the complexity of the plant breeder's work, of the different skills that he must practice and of the judgements that he must make, that while genetics and geneticists may serve him, and serve him well, they will not replace him.

References (Mather)

Biffen, R. H. 1906 An attempt to find a basis for the improvement of the barley crop. *J. Inst. Brewing* **12**, 344–369.
Biffen, R. H. 1907 Studies in the inheritance of disease resistance. *J. agric. Sci., Camb.* **3**, 109–128.
Darlington, C. D. 1932 *Recent advances in cytology*, 1st edn. London: Churchill.
Darlington, C. D. 1939 *The evolution of genetic systems*. Cambridge University Press.
Pandey, K. K. 1975 Sexual transfer of specific genes without gametic fusion. *Nature, Lond.* **256**, 310–313.
Pandey, K. K. 1978 Gametic gene transfer in *Nicotiana* by means of irradiated pollen. *Genetica* **49**, 53–69.
Rees, H. 1961 Genotypic control of chromosome form and behaviour. *Bot. Rev.* **27**, 288–318.
Shepard, J., Bidney, D. & Shahin, E. 1980 Potato protoplasts in crop improvement. *Science, N.Y.* **208**, 17–24.
Virk, D. S., Dhahi, S. J. & Brumpton, R. J. 1977 Matromorphy in *Nicotiana rustica*. *Heredity, Lond.* **39**, 287–295.
Wernicke, W. & Brettell, R. 1980 Somatic embryogenesis from *Sorghum bicolor* leaves. *Nature, Lond.* **287**, 138–139.